Ori. * * ○ * Occ.

Ori. ○ * * * Occ.

Ori. * * ○ Occ.

SYSTEM

INFRASTRUCTURES SERIES

edited by Geoffrey C. Bowker and Paul N. Edwards

Paul N. Edwards, *A Vast Machine: Computer Models, Climate Data, and the Politics of Global Warming*

Lawrence M. Busch, *Standards: Recipes for Reality*

Lisa Gitelman, ed., *"Raw Data" Is an Oxymoron*

Finn Brunton, *Spam: A Shadow History of the Internet*

Nil Disco and Eda Kranakis, eds., *Cosmopolitan Commons: Sharing Resources and Risks across Borders*

Casper Bruun Jensen and Brit Ross Winthereik, *Monitoring Movements in Development Aid: Recursive Partnerships and Infrastructures*

James Leach and Lee Wilson, eds., *Subversion, Conversion, Development: Cross-Cultural Knowledge Exchange and the Politics of Design*

Olga Kuchinskaya, *The Politics of Invisibility: Public Knowledge about Radiation Health Effects after Chernobyl*

Ashley Carse, *Beyond the Big Ditch: Politics, Ecology, and Infrastructure at the Panama Canal*

Alexander Klose, translated by Charles Marcrum II, *The Container Principle: How a Box Changes the Way We Think*

Eric T. Meyer and Ralph Schroeder, *Knowledge Machines: Digital Transformations of the Sciences and Humanities*

Geoffrey C. Bowker, Stefan Timmermans, Adele E. Clarke, and Ellen Balka, eds., *Boundary Objects and Beyond: Working with Leigh Star*

Clifford Siskin, *System: The Shaping of Modern Knowledge*

SYSTEM

Ori. * * ○ * Occ.

The Shaping of Modern Knowledge

CLIFFORD SISKIN

The MIT Press
Cambridge, Massachusetts
London, England

This book was set in Bembo by Toppan Best-set Premedia Limited. Printed and bound in the United States of America.

Library of Congress Cataloging-in-Publication Data

Names: Siskin, Clifford, author.
Title: System : the shaping of modern knowledge / Clifford Siskin.
Other titles: Shaping of modern knowledge
Description: Cambridge, MA : MIT Press, 2016. | Series: Infrastructures | Includes bibliographical references and index.
Identifiers: LCCN 2016015626 | ISBN 9780262035316 (hardcover : alk. paper)
Subjects: LCSH: Interdisciplinary approach to knowledge. | Knowledge, Theory of. | System theory. | Galilei, Galileo, 1564-1642--Knowledge--Science.
Classification: LCC BD255 .S57 2016 | DDC 003--dc23 LC record available at https://lccn.loc.gov/2016015626

Parts of the argument in chapter 4 appeared in an earlier form in "The Year of the System," in *1798*, ed. Richard Cronin (Basingstoke: Macmillan, 1998), 9–31. Parts of the argument in chapter 5 appeared in an earlier form in "Novels and Systems," *Novel* 34, no. 2 (spring 2001): 202–215. Parts of the argument in chapter 6 appeared in an earlier form in "William Wordsworth" in *The Oxford Encyclopedia of British Literature*, vol. 5, ed. David Kastan (Oxford: Oxford University Press, 2006), 326–334) and in "The Problem of Periodization: Enlightenment, Romanticism, and the Fate of System," in *The Cambridge History of English Romantic Literature*, ed. James Chandler (Cambridge, Cambridge University Press, 2009), 101–126.

10 9 8 7 6 5 4 3 2 1

For
Michael Sprinker
and
Robert Cummings

CONTENTS

ACKNOWLEDGMENTS

For close to a decade, I have been collaborating with colleagues on a start-up. The Re:Enlightenment Project joins institutions and individuals across disciplines and professions who share a common purpose: reconceiving how knowledge works in the world. It makes a future for Enlightenment by better understanding the ways in which Enlightenment made us.

To that end, the project has formed collaborative networks through international exchanges, experimented in new forms of knowledge conceived and conducted by those networks, and disseminated the results in articles, edited collections, and monographs. All of those outputs have shaped each other. Each one is a collaborative product of our joint effort to build a new infrastructure for knowledge work.

It is particularly appropriate, then, that what began as an essay solicited for a collection on a canonical moment in literary study has scaled up into a book in the Infrastructure series from a press committed "to exploring new fields and new modes of inquiry." I hope it contributes to that series not only by explaining the relationship between system and infrastructure but also by exemplifying why infrastructure such as the Re:Enlightenment Project matters. I thank the series editors, Geoffrey C. Bowker and Paul N. Edwards, my editor at the MIT Press, Katie Helke, and my manuscript editor, Judith Feldmann, for so ably enabling this match between content and the way it makes its way into the world.

Leslie Santee Siskin has been my match in every way that matters. She also gave the Re:Enlightenment Project its name and—through her expertise in the organization of knowledge and diplomatic skill in negotiating it—has been central to all of our efforts to make it live up to that name. We have also collaborated for a long time on another infrastructural start-up—the fundamental one of family—which has grown faster than this book. My thanks and love to Cri/Raf/Dylan/Henry, Nat/Sarah/Alex, and Johanna/Chris/Caden/Devin.

William Warner and Peter de Bolla are my coauthors in everything I do—even when we are not writing directly together. Bill and I have worked in tandem closely enough and long enough to be pegged in print as a "Wordsworth and Coleridge"—yet another "male double act" (Russell 2013, 78). I have thoroughly enjoyed our performances, especially constructing the argument for *This Is Enlightenment*—a volume crucial to the success of Re:Enlightenment and to my realizing what a book about system could—and should—do. To work with Pete is to await the sudden synthesis, the moment when an idea for a new format reanimates the project's efforts at exchange or when a new method does what a method is supposed to do—provide a "way forward." Working with Pete in Cambridge as a Leverhulme Visiting Professor enabled a smooth transition from finishing *System* to Re:Enlightenment's next venture into digital knowledge.

I was introduced to Bill by John Bender, whose long and dear friendship has been a constant source of support (and criticism) during my entire career, and especially at crucial moments in the writing of this book. Anne Mellor has also been a constant source of friendship and advice, as we jointly pursued another kind of knowledge work: coediting a series for Palgrave-Macmillan. My thanks as well to a number of other long-term friends and colleagues: Nancy Armstrong, Len Tennenhouse, Herbert Lindenberger, Mary Poovey, Michael Hill, Devoney Looser, Philip Martin, and Robert Miles.

Two invitations from Richard Cronin changed my work and my life. He invited the essay that became my first work on system—and he invited me to the University of Glasgow as well. My heart is and always will be in Scotland and with the friends I made there over the years, including Richard, Dorothy Macmillan, Drummond Bone, Nicola Trott, Jan Todd, Murray Pittock, and Stuart Gillespie. At New York University, I have enjoyed teaching and working with Bill Blake, Lisa Gitelman, and Paula McDowell; Chris Cannon has been the best kind of chair—an enabling one. During my stints at Cambridge, I have enjoyed friendship and advice from James Raven, Simon Schaffer, and Chris Prendergast.

Simon Goldhill, director of Cambridge's Centre for Research in the Arts, Social Sciences and Humanities, helped me to secure the Leverhulme and teach the Mellon Experimental Concept Lab with Peter de Bolla, and he organized an intense and very useful seminar on my work on system.

That work has gone hand-in-hand with the work of the Re:Enlightenment Project. I thank my cofounder, Kevin Brine—an Enlightenment figure in his remarkable blend of intellectual firepower and business acumen—for those white-hot moments in which we conceived of and launched the project. Thanks as well to the other members of the original steering committee: book historian and publisher William St. Clair, curator of the British Museum's Enlightenment Gallery, Kim Sloan, Peter de Bolla, Leslie Santee Siskin, William Warner, Robert Young, and, more recently, Murray Pittock and Helge Jordheim. Helge and Murray have been among my best interlocutors over the past few years.

So too has the physicist David Deutsch, the father of the quantum computer and one of our most important scholars of Enlightenment and the history of science. In his books, in conversation, in the seminar room with my students, and in Re:Enlightenment events, David's explanations of explanatory knowledge are game changing. His centering of the terms "guesswork" and "error correction" proved crucial to my figuring out the historical and epistemological role of system in Newton and Newtonianism. And his current work plays a key role in my turn to the future in the Coda to this book.

Finally, and crucially, I acknowledge and highlight the significant contributions my research assistants and my doctoral students at Stony Brook, Glasgow, Columbia, and NYU have made to this book. In particular, Anthony Jarrells and Tom Mole did important archival work on system. Tony found wonderful references to system in a wide range of late eighteenth- and early nineteenth-century genres. Tom accumulated titles of systems from the same period to provide a portrait of what he called "a new kind of citizen, Systematic Man" (see chapter 4). Yohei Igarashi researched the intersection of system and blame and later teamed with Seth Rudy to work on *This Is Enlightenment* and the Re:Enlightenment Project. Rachael King succeeded them as an invaluable assistant for—and contributor to—the project and my final push to finish *System*. Laura Yoder provided a helpful close reading of the manuscript.

Mark Algee-Hewitt, my former student and now colleague in the North American Concept Lab, designed the computational environment and visualization that I call "Tectonics." There are few other scholarly activities more pleasurable than playing intellectual leapfrog with those we mentor. Mark picked up on my classroom suggestion that we could make more sense of the history of the sublime by counting, and then transformed himself within the next few years into a pioneer in digital inquiry. I asked him to explain in his own words what Tectonics does and how. You can find his explanation in appendix A.

I lost two of my closest personal and intellectual friends during the writing of this book: Michael Sprinker at the start and Robert Cummings as I finished it. If you knew Michael, you learned what a knowledge worker could be and why being one matters. If you knew Bob, you learned what it means to be learned—and profoundly generous. This book is dedicated to them.

New York City, 2016

PROLOGUE: "THE MOST PRIMITIVE QUESTION"

> We have discovered four wandering stars, known or observed by no one before us. … We will say more in our System of the World.
> —Galileo Galilei, *Sidereus Nuncius* (Starry Messenger), 1610

> I think we can ask the most primitive question: What is a system? Now I don't think we can answer it. It's a sort of undefinable. You start with the idea of a system.
> —Professor of Physics Leonard Susskind, Lecture I of "The Theoretical Minimum," an online introduction to quantum mechanics, Stanford University, 2012[1]

For over four hundred years, we have been saying, as Galileo promised in his message from the stars, more and more about system. But saying more, as Leonard Susskind declares at the start of his introductory lecture on quantum mechanics, has not helped us to say what it is. In fact, the more we say about system—the more we use it in different contexts and to different ends—the less likely we are to answer the "what" question with a definition. What we can do instead with an "undefinable" is pay very careful attention to what *kind of thing* we assume it to be. If this book could send its own message across the disciplinary divide, the first point would be quite simple: don't "start with the *idea* of a system" (emphasis mine) and get stuck with the problem of a fixed definition; start with system as a particular kind of thing that can be explained. Our word for *kind* is *genre*.[2]

By engaging system as a genre—as a form that works physically in the world to mediate our efforts to know it—this book illuminates system's role in the shaping and reshaping of modern knowledge. To take the idea route is to lose sight of that role as we sink not only into the problem of definition but also into the familiar divide of the abstract versus the physical. Understood as a genre, however, system can describe what we see, as with Galileo, operate a computer, or be made on a page, like a sonnet or a letter. Starting in the seventeenth century, more and more people wrote and published works that they named and titled "system"—turning system into one of the genres, along with competitors such as treatises and essays, that filled the Enlightenment with the work of writing. These genres competed because they shared features with each other; each one is discernible as a kind by the features it has and has not shared with other kinds. In that sense, genres exist in their interrelations with other genres.

I identify features of system, such as its scalability, that can help us explain why system came to play such a central role in efforts to know the world for so long. Beginning with Galileo's sighting of Jupiter's lunar system, the argument tracks what people have called "system" in its many intellectual and social incarnations, from Newton's system and the proliferating systems that generated Enlightenment, to the modern disciplines that emerged from it, to Darwin's algorithmic system of survival and our own plethora of new uses for, and kinds of, system—including network, nervous, computing, and communication systems, as well as systems theory, self-organizing systems, and system professionals.

Because the question of system is—as Susskind sees it in physics—a "primitive" (fundamental) one in so many of those realms, it deserves answers that can speak across the centuries and across the differences. That is what I hope my turn to genre can deliver. Since, as I invoke them here and explain in detail in chapter 1, genres are dynamic—they are historically emergent classifications—those answers will be robust not essentialistic, empirical not logical. What things at what times were actually called system? What were their characteristic features? Did those features change? Were some more persistent than others? Were those features shared with other genres? How did those combinations of features change? What work did these genres perform? And with what effects?

By altering how we know the world, system has changed it. But seeing it as a genre highlights how system itself has also changed. I will examine that versatility in terms of system being both scalable (systems within systems) and adaptable (to different conditions and substrates). The plethora today of new uses for, and kinds of, system emerges from, or is made visible by, the shift into electronic and digital technologies. In the eighteenth century, systems were also powerful and ubiquitous. However, the technology that embodied them then was different: the printed word was the proliferating technology, and system was increasingly invoked and used as a specific form of writing. That form competed with other written forms, particularly the essay, to great effect, helping to reshape and reorganize knowledge into the modern disciplines.

On one horizon of this book, then, are earlier, pre-Enlightenment efforts to explain the world by articulating what Galileo called a "system of the world." On the other is our post-Enlightenment world full of knowledge systems, a world in which system may have yet one more disorienting explanation up its sleeve. In what some are proposing as another Baconian instauration—a revolutionary new kind of science—the Enlightenment's once startling conviction that the world can be known by making systems is being turned on its head. In consensus-making collections such as *A Computable Universe: Understanding and Exploring Nature as Computation* (Zenil 2013), knowledge itself—embodied in a computational system—makes the world.

To grasp what's at stake in scanning these horizons—that is, to understand system's purchase on the modernity we now inhabit—we need to engage system as not just an abstract concept or idea but as something materially *in* the world—something specific, concrete, and countable. That is why system has worked in that world to shape knowledge—and that is why I am counting on genre to help us identify and understand the history of that work. This is not, of course, the conventional way to discuss system or track it over time. In fact, we tend not to dwell on system as something that has a history or, per Susskind, even a definition; most often, we either make systems or we use them. Prior engagements with system have almost always been a matter of imagining or engineering a better one—or claiming, often through systems theory, that things "actually" work that way.

When they don't, system takes the blame. Today, even as we admit new systems of all kinds—operating systems, support systems, ecosystems, phone systems—into our lives, we do not hesitate to "blame the system," as in that which, in its most popular form, works both too well—"you can't beat The System"—and not well enough—it always seems to "break down." But what is surprising is that this is not just a recent habit—a bad attitude of cynical times. System and blame, I will demonstrate, have tales to tell about each other. When did systems become things that *could* be blamed? When did those things become means for reshaping knowledge? And when and how did those means manifest themselves in the world as social, political, and cultural forms?

FINDING ANSWERS

The endeavor of this book is to make system's many manifestations visible, and thus magnification is as crucial to my message as it was to Galileo when he aimed his telescope toward Jupiter. In both cases, the point is to increase that "wonderful effect" of making distant things appear "as though nearby" (Galilei [1610] 1989, 37). For Galileo, as I detail below, this was a matter both of optimizing his optics and of shaping a message from the stars that could best convey and amplify what he saw.

This book begins by taking on those same two tasks. After chapter 1 introduces genre as a better way to see system, chapter 2 turns to devising a form of inquiry that can best convey system's message. The most likely suspect, of course, is to write "a history" of system. But among this book's unexpected findings is that one of the most telling ways that system shaped modern knowledge was to reshape history itself. *Historia* has taken many forms; like system, it has been a kind of writing that has changed over time. Those changes matter for this book because of system's role in changing history—and also because history, in turn, played a role in changing system: the two genres transformed each other.

I track their fates by first recovering the roles that Galileo and Francis Bacon assigned to system and history in their plans for advancing knowledge. In doing so, I recover from Bacon a form that I hope advances my effort to know system. The name of that form—*literary history*—may appear familiar, but my strategy here is to revamp the scope and purpose of our modern version of it. I take my cue from Bacon, who saw the writing of literary history as crucially important to the advancement of knowledge—and for a simple reason: its subject in his scheme was knowledge itself. "*Historia Literarum*" was the living "eye" of the "history of the world," for it recounted that world's "story of learning"

(Bacon [1605] 2008, 175–176).[3] For Bacon—and through much of the eighteenth century—*literary* was a comprehensive term inclusive of all written records, not the selection of specific kinds and qualities that became our modern disciplinary category of "Literature."

To reclaim this earlier and broader sense of *literary* is to acknowledge our own place in the history of how, as Bacon put it, "knowledges" change. Our moment of change is marked by the institutional configuration of narrow-but-deep disciplines (Siskin 1998a, 20) and by increasing attention paid to the possibility of alternative arrangements (e.g., inter- and dedisciplinarity). This book participates in that moment by taking a form of inquiry out of its current disciplinary silo so that it can track system's role in the shaping of those silos. My strategy, then, is to recover and take on Bacon's unrealized ambition for literary history *as the story of learning* in order to tell the specific tale of system's role in it.

Bacon saw the difficulty of assembling a literary history at this imposing scale, so he started with a wish list of 130 "particular" histories to help him to do it (Bacon [1620] 2000, 232–238). I describe that strategy in chapter 2 and use it to identify histories particularly useful to this book. These are histories in which system's effects on knowledge are most visible—histories of mediation, blame, and the real. I hope these moves—recovering the force of both Bacon's ambition and *historia*'s robust record of operating in the plural—clarify rather than cloud. Not surprisingly in a book about a form, form matters—and these experiments in form aspire to be useful methodologically for others venturing into projects of this scope.

SYSTEMS ABOVE/SYSTEMS BELOW

To focus on how knowledge—including my own—takes form, we need to zoom not only up and out to the celestial systems of Galilean cosmology but also down and into the worldly systems of what we currently call "infrastructure." Since the early twentieth century, that term has pointed us to what is "infra," that is, "below." Although aimed in a different direction than Galileo's telescope, the analytic tool of infrastructure also magnifies. Through it, we see how every society relies on underlying structures to bring it to life—structures supplying such services as transportation, energy, and communication. And infrastructure directs our attention as well to the next level down—to how those structures depend in turn on the specific physical forms that support assumes: roads and bridges, power grids, telephone networks.

Most important for this book, however, is the distinctive way that infrastructure achieves its highest resolution. The workings of infrastructures become most visible when infrastructures don't work—when, in Paul Edwards's words, "the server is down, the bridge washes out, there is a power blackout" (Edwards 2010, 9). Only then, when we try to fix them, do we fully appreciate how surprisingly complex they are. Rebuilding requires not only silicon, concrete, and wires but the information and knowledge that built—and are built into—the structures themselves.

To attend to infrastructure is thus to attend to the subject Bacon specified for his "just story of learning": "knowledges … their diverse administrations and managings" (Bacon [1605] 2008, 176). My contribution to that story is thus also a contribution to MIT's Infrastructures series. Being a part of that series offers a broad disciplinary context and audience for my focus on how knowledge is generated, shaped, and put to work in the world. In turn, this book offers a specific contribution to the study of infrastructure—and for a specific reason: work on infrastructure has posed the very same "primitive question" that troubles physics. The fact that the question surfaces here as well confirms how fundamentally primitive a question it is: What is system?

System is not infrastructure. The system question arises in this field not because infrastructures are systems but because system is their analytic foil; it is invoked to highlight the distinctive nature of infrastructures. Because they are "sunk into, inside of, other structures, social arrangements, and technologies," infrastructures are never "built from the center with a single design philosophy." Rather they are "built from the ground up in modular units" (Edwards et al. 2007, 33), their development an "oscillation between the desire for smooth, system-like behavior and the need to combine capabilities no single system can provide" (Edwards 2010, 12).

System, then, is a crucial component of infrastructures but only one of the units that comprise them—the "spectrum" running from systems to networks to Internetworks or webs. Making these jumps in form is thus a central issue for infrastructure studies: how do infrastructures evolve in scope and complexity from single systems to "widely shared, highly accessible" compound structures (Edwards et al., 2007, 12, i)? At stake here are the dynamics of scale, specifically the scalability of systems. "How did "the world" become a *system*?" asks Paul Edwards at the start of his account of climate science as a "global knowledge infrastructure" (Edwards 2010, 8, 3).

That is precisely the same question that drives my inquiry into system's emergence as the primary form of Enlightenment and the norm for modern knowledge. *System* offers to this series an account of scalability as one of the primary features of the genre of system. It can thus document over time system's role in scaling knowledge up and down from Galileo's "System of the World" to Newton's to Enlightenment encyclopedism to modern disciplinarity and, now, the computational universe. By aspiring to the scope of a Baconian literary history, we can better explain how these varied uses of system shaped knowledge in a manner that supports the building of infrastructures. Conversely, we can also grasp how, because of scalability, infrastructures can be subsumed into larger systems. The fact that infrastructures are not systems is precisely the reason we need to know what system has been, what it is, and what it can enable.

As an example of long-form knowledge in print, this explanation is itself, of course, a kind of system; in writing a book, I have made the modern version of what we will see is an eighteenth-century choice—one enshrined in Samuel Johnson's *A Dictionary of the English Language* (1755)—between writing a system and writing an essay. And as a system, the feature of boundaries—of what should and can be within the bounds of this book—comes to the fore. That is a concern of every book, as it is for every system, but it is magnified here by Bacon's vision of a comprehensive literary history stretching from "age to age." That full stretch cannot, of course, be the task of one book. The criteria for what will have to find a place in future systems, as well as links to earlier efforts by others to scale up, are taken up more fully in the discussion in chapter 2 of histories for system.

For now, I want to foreground two protocols that governed my decisions about what to include in this book. By *protocols* I mean enabling constraints. This book is neither an example of "systems theory" nor a survey of every instance of system and the theorizing of it. It's not an example, because my goal is not to theorize system as an explanation of the way things actually work. It's not a register of every appearance and use of system since the early seventeenth century (though it features many of them), because my goal is to explain, not survey. This book mounts a specific, historical argument *about* system: it is an explanation of how system became a primary form for shaping knowledge during the Enlightenment and where it might be headed in the future. My hope for this book is that its message of the significance of system and its staying power—that there was and will be, as Galileo put it, more to say[4]—resonates with the message he sent as system first entered into the burgeoning world of knowledge making in the early seventeenth century.

A MESSAGE FROM THE STARS (SIDEREUS NUNCIUS), 1610

Ori. * * O * Occ.

> On the seventh day of January of the present year, 1610, at the first hour of the following night, when I was viewing the constellations of the heavens through a spyglass, the planet Jupiter presented itself to my view. As I had prepared for myself a very excellent instrument, I noticed a circumstance which I had never been able to notice before, owing to want of power in my other spyglass. That is, three little stars, small but very bright, were near the planet. Although I believed them to belong to the number of fixed stars, yet they made me wonder somewhat, because they seemed to be arranged exactly in a straight line parallel to the ecliptic, and to be brighter than the rest of the stars equal to them in magnitude. … I therefore concluded, and decided unhesitatingly, that there were three stars in the heavens moving around Jupiter, like Venus and Mercury around the sun. This was finally established as clear as daylight by numerous other subsequent observations.
>
> —Galileo Galilei[5]

The message delivered in 1610 was clear—"as clear as daylight," as Galileo exclaimed after gazing into the night sky. Stars that he had believed to be "fixed" turned out to be, upon subsequent observation, "moving around Jupiter." These three little stars, he "decided unhesitatingly," were actually three moons. But that wasn't why Galileo decided to rush his "new observations" into print as "an announcement to all philosophers and mathematicians."[6] The discovery of new heavenly bodies was certainly exciting, but that wasn't the "message."

What mattered more than the bodies themselves was what they formed: together with their home planet, these bodies constituted a *system*. And what was at stake for knowledge—knowledge of the "true and physical constitution of the world" (Galilei 2008, 152)—was not that this was the first but the *second* such system that had been observed. To the earth and its moon, Galileo had added a second system—and that made all the difference. It changed how the first was understood. If all heavenly bodies revolved around the earth, then the moon was like the sun and the planets—just another part of a single and comprehensive "system of the world." But seeing Jupiter as a system with its own moons cast the earth and its moon in a different light: as a system unto themselves, the moon assumed a new status as the *only* body to revolve around the earth.

Copernicus had, of course, made that argument in *De revolutionibus orbium coelestium* in 1543, but for seventy years, the debate had stalled, sharing the same fate that Galileo's contemporary, Francis Bacon, declared for all knowledge. "The sciences," Bacon complained in arguments for "advancement" and "restoration" that chronologically bracketed Galileo's message, "are almost stopped in their tracks."[7] Although some astronomers were attracted by its mathematics, heliocentrism had failed to provoke even the official ire of the Church through the remainder of the sixteenth century.

The basic reason for what looks to us like a puzzling delay was that there was too much to lose. "Accepting Copernicus's system," as Albert Van Helden points out, "meant abandoning Aristotelian physics," and thus opening up a host of questions to a need for entirely new explanations:

> Why does a stone thrown up come straight down if the Earth underneath it is rotating rapidly to the east? Since bodies can only have one sort of motion at a time, how can the Earth have several? *And if the Earth is a planet, why should it be the only planet with a moon* [emphasis mine]?[8]

To discover that it was not the only one, Galileo realized, was to break the logjam—to acquire, at last,

> a notable and splendid argument to remove the scruple of those who can tolerate the revolution of the planets around the sun in the Copernican system, but are so disturbed by the motion of one moon around the earth (while both accomplish an orbit of a year's length around the sun) that they think this

> constitution of the universe must be rejected as impossible. (Galilei [1610] 2008, 83)

At stake in sorting out "system" was not just knowledge but knowledge that was load bearing[9]—knowledge, that is, that made more knowledge possible. The translator's choice of "Splendid" is thus not quite adequate to Galileo's Latin adjective for this "argument": "praeclarum" points to a specific kind of argument, one that brings clarity.[10] This argument made new knowledge possible, preparing the way for the resolution of the revolution debate in Galileo's *Dialogue on the Two Chief World Systems* two decades later (1632).

The message of the stars in 1610 was thus clear and enabling. Galileo saw that those stars were moons, and what they told him was something new: the world is a system full of systems. This book—the one under your observation—is also about seeing systems; in a sense, it is an update on their proliferation. For us, the system of the world has become a world full of systems. In both tellings, systems are found to be pervasive, and finding more of them—especially in places where they have not been observed before—raises possibilities for new knowledge. That same year, for example, in which Galileo found a "system" in deep space—1610—is the year the *Oxford English Dictionary* (*OED*) cites for the first appearance of the word *system* in English to refer to "The whole scheme of created things, the universe":

> 1610 J. Selden Michael! in M. Drayton Poems (rev. ed.) sig. A5, Thy Martiall Pyrrhique, and thy Epique straine Digesting Warres with heart-vniting Loues. (The two first Authors of what is compos'd In this round Systeme All).[11]

Why does system surface in these different ways—appearing in Galileo's glass and venturing into the vernacular? Why then? Why does this apparent coincidence expand so rapidly into a historical cluster over the next few decades? By the 1630s, when Galileo publishes his "World Systems" manifesto, *system* becomes a descriptor for things of all kinds, as in the *OED*'s second sighting of system in 1638: J. Mede, "Man's life is a systeme of divers ages. … The yeare is a systeme of foure seasons."

THE SHAPE OF SYSTEM AND THE SHAPE OF THIS BOOK

With the help of new tools—but tools that look through time rather than into space—I hope to track this proliferation, including its causes and its effects. To that end, I have organized this book into three parts. They link this monograph to larger, collaborative endeavors that have helped me at crucial points to reenvision it, from the making of the collection *This Is Enlightenment*, to the ongoing efforts of the Re:Enlightenment Project (www.reenlightenment.org), to understand and to transform our Enlightenment inheritance. In addition to its other goals, this book aims to contribute to that joint enterprise. The premise behind my title is that engaging system is crucial to understanding Enlightenment and the forms of modern knowledge that emerged from it. And that understanding is, in turn, essential to reshaping the legacy of Enlightenment today. To that end, I have organized *System: The Shaping of Modern Knowledge* so that the woof that threads through its chronological warp follows the project's touchstones for recovering and reanimating the Enlightenment goal to know the world:

Those touchstones are:

- *Past and Present*—How should the historical Enlightenment relate to the possibilities for Enlightenment now?
- *Mediating Technologies*—How can we reconceive our structures and tools for making knowledge work in the world?
- *Connectivities*—How can we reconnect across the boundaries and divisions—institutional, professional, disciplinary—that arose after Enlightenment?

To focus in on system, I have recast these three queries to serve as coordinates for exploring that genre: What is system, and how has it changed? How has system mediated knowledge? How has system worked in the world? Configured by these retouched touchstones, this book stands not only as a stand-alone monograph but—playing on what I call system's scalability (wholes that become parts of larger wholes)—as a project that is part of a larger collaborative Project.

With all the risks of omission and distortion that a distillation of a book carries, here is a bullet-point list, paced by changing deployments of system, of the principal arguments of this book. To present them in this format—the sequential schematizing of parts in a whole that, I show, features in so many systems—highlights how central the travels of system still are to writing the long-form knowledge that we call "books."

- Part I, "Past and Present," sets the book's formal strategy and its temporal horizons in the ways I have already described. To recuperate literary history as what Bacon called the story of learning, I recover the multiplicity of *historia* itself. Instead of writing a single "history of system," I build my argument by putting system *into* a range of different histories. The other significant innovation in the first two chapters is the debut of what Bacon would call a new "instrument." Like Galileo's spyglass, the computational and visualization technique I call Tectonics zooms in—in this case, to clarify how our modern disciplines emerged from Enlightenment through the late eighteenth-century collision of two genres: system and history.
- Part II, "Mediating Knowledge," offers a new take on the history of science by detailing how the turn from Scholasticism in the seventeenth and eighteenth centuries took the form of a gradual turn to system as the "firmer" form of what the physicist and Enlightenment scholar David Deutsch (2011) calls "guesswork" (10). Deutsch's explanation of science as itself explanatory rather than descriptive has helped me to recast system as a historically specific genre of choice for explanation. That turn to system as an enabling form of guesswork was completed in grand fashion with Newton's decision to communicate his principles and laws philosophically by adding a "System of the World" to his treatise. In chapter 3 I explain how and why Newton came very close—repeatedly—to sending the *Principia* into the world without any "System" at all. Why was system such a vexed issue in the late seventeenth century? What was at stake for Newton in choosing to write a system? And why, once it did make it onto the printed page, did *system* become so successful that a copy of Newton's system was launched into space as one of humanity's calling cards three centuries later? How did that particular form of knowledge come to represent our species' ability to, as Francis Bacon put it, "advance knowledge"? The primary generic marker of what came to be called Enlightenment, I conclude, was the monumental effort to scale up systems into Master Systems that persisted from roughly the 1730s through the 1780s.[12] During the next two decades, I argue in chapter 4, those efforts at comprehensive mastery gave way to different uses of system—to delimited and dedicated systems and to the dispersing of systems into other forms, including the specialized essays of the modern disciplines. Their "travel" filled the world in new ways.

- Part III, "Connectivities," pursues the effects of that saturation by identifying the social incarnations of system—the ways that system re-formed society itself. Turning to system as a genre interrelated with other genres provides us with a way of understanding and articulating how behaving in writing—and in other technologies and substrates—connects to other kinds of social behaviors. For this touchstone, I engage system as a form that mediated modernity, thus highlighting connectivities as well as differences between past and present. I show how that mediation gave rise to the concept and practice of systematizing and its sister phenomenon of "instituting." Each of the three chapters in this third part focuses on one of the three categories through which we have come to apprehend ourselves and our behaviors: the political (including the logic of liberalism), the cultural (and the formation of Literature), and the social (from clubs to interfaces). Together they describe one of the primary, cumulative effects of system's work in the world: how we came to blame The System.

What follows in the Coda to this book is necessarily more speculative, for it brings us into the open-endedness of the present. But at the very least, I hope to show in conclusion that there is considerable evidence that now may be the time of one more bullet point in system's shaping of knowledge and ourselves.

Part I PAST AND PRESENT

FROM THE "SYSTEM OF THE WORLD" TO A WORLD FULL OF SYSTEMS

1 ENGAGING SYSTEM

> The notion of a philosophical "system" or of philosophical "systems" is so well established today that it is hard for us to believe that it has a history at all.
> —Walter Ong, 1956

SIGHTING SYSTEMS "WITH THE HELP OF A SPYGLASS LATELY DEVISED"

The graph in figure 1.1 is a way of seeing. It brings into focus the subject of this book, system, in a manner, and using technologies, only lately devised.[1] Its message is remarkably clear: the astonishing—and astonishingly linear—rise of references in print to "system" during the eighteenth century in Britain. Not only does system have a history, but it is a particularly memorable one once we visualize it. Explaining this takeoff, and thus system's central role in Enlightenment and its aftermath, is a core agenda of this project. To accomplish it, I'll follow Galileo's lead on the title page of his message and begin with tools—both the tools that enabled his sighting and those I will use for mine. For his spatial turn to the heavens, this acknowledgment foregrounded for his readers that what we see depends on how we see. My turn to the temporality of system adds an ironic twist: seeing systems from the past depends on deploying new systems (algorithmic, digital) in the present.

In a typeface only slightly smaller than the one used for his own name, Galileo centered on that page his debt to the tool that helped him to discover the "great and wonderful sights" "unfolding" within (Galilei [1610] 1989, 26–27). As a tool of his own devising, this "perspicilli" was both enabling for and enabled by him. In the formulation later made famous

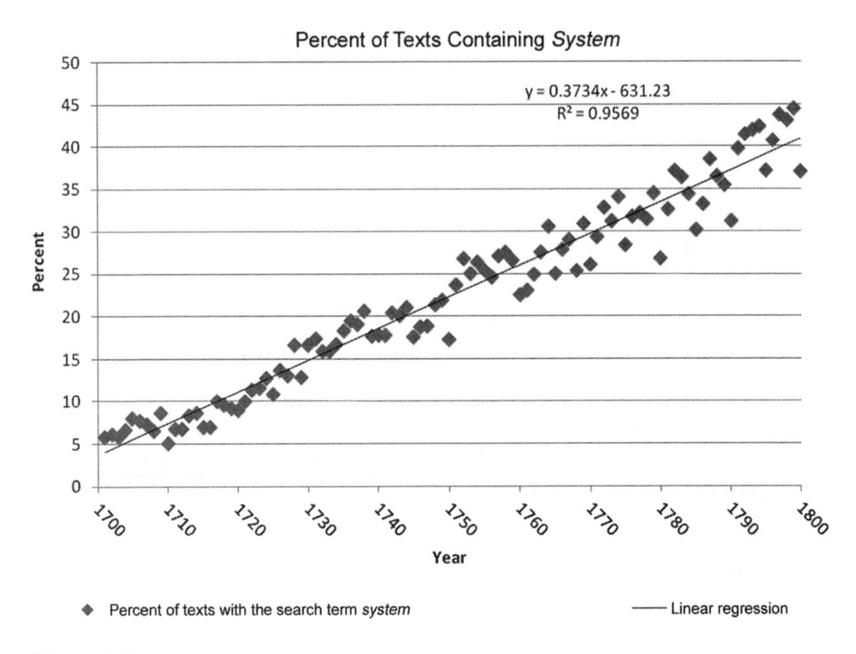

Figure 1.1
Using *Eighteenth Century Collections Online* (see appendix A).

by Marshall McLuhan, the medium of the "glass" he invented became an extension of his own senses.[2]

That Galileo understood his relationship to his tools in precisely these terms is evident in his courteous treatment of the philosophers whose work he was upending. Instead of dismissing Aristotle for being wrong, Galileo emphasized that his predecessor was simply not "in possession of present day sensory observations." Since Aristotle gave "first place" to "plain sense experience," argued Galileo, it is "very reasonable to believe that if the senses had shown to him what they have shown to us, he would have followed the contrary opinion, to which we are now led by such marvelous discoveries" (Galilei [1613] 2008, *History and Demonstrations Concerning Sunspots*, 100).

Such generosity, born out of a sense of what we would now call historicity, assumes that retooling can produce historically different ways of knowing. Galileo shared with his contemporary Bacon the strategy of using this notion of change to choose, in Bacon's words, the "happy" option of "carry[ing] out our design without touching or diminishing the honour and reverence due to the ancients." New tools, Bacon argued, opened a "way to the intellect" that the ancients simply did not have. Ancient versus modern was thus not a "contest" but history—a history that told of the good "fortune" in living in a moment of different "resources" than they did (Bacon [1620] 2000, 29).

Bacon famously specified those resources in 1620 as three "mechanical things" with recent origins. Together, Bacon argued, they "have changed the face and condition of things all over the globe": "the art of printing, gunpowder and the nautical compass." More than "soil, climate or bodily qualities," these things, Bacon believed, made the difference he perceived between "civilized" Europe and "barbarous" New India. In his history, the most powerful "empire" had less purchase on the present than this trio of tools (Bacon [1620] 2000, 100).

What was distinctive about Bacon's formulation, however, was not this trio. Almost thirty years earlier, in 1591, the infamous astrologer and physician Simon Forman had already listed "the Compasse," "gunne powder," and "the Arte of Printing" as "things very needfull and necessary which are now vsed" (Forman 1591, image 6). In Bacon's version, however, these mechanical mediations—things that Man could use to act on Nature—shared the new toolbox with intellectual mediations. Those "resources" were instruments of "control" that "open[ed]" a "new and certain road for the mind." They were not physical things but rules for that "road": a "method" to supplant Aristotelian logic. They were needed, he argued, because that old organon—the collective title traditionally given to Aristotle's logic—only "keeps and accumulates" "faulty" things. The syllogisms and dialectic of Scholasticism produce "a kind of giddiness, a perpetual agitation and going in a circle." For Bacon, *method* was a "machine" for getting things under control—for getting the mind's "business done" (Bacon [1620] 2000, 2, 28).

Combining both kinds of resources was as crucial for Galileo as it was for Bacon. Seeing the stars for himself was only the first step in the business of conveying their message. The second was to prepare his observations for "inspection and contemplation" by others. After acknowledging the mechanical "instrument" that made the stars "manifest" to his gaze—a "new spyglass"[3]—he placed his discoveries into a series of explanatory frames: how they "add" to what is visible, how they "please" the "eye," how they "end" "debate," and, especially, how they "give notice" that there are "stars" wandering around "stars."[4] Just as his telescope brought the stars themselves near, so these frames methodically magnified their message, eventually revealing a claim that "greatly exceed[s] all admiration"—the claim about systems within the system of the world (Galilei [1610] 1989, 35–36).

Galileo did not overestimate the effect of his discovery. It has earned him a place in the pantheon of Western knowledge as, in Albert Einstein's words, "the father of modern physics—indeed, of modern science altogether." He deserves that accolade, argued Einstein, because he was the first to insist that "pure logical thinking cannot yield us any knowledge of the empirical world" (Einstein 1954, 271). My point here is not to reanoint Galileo as "the father" or "the first," but to clarify why he is understood to have played a key role in the transition from the old to the new organon—what Bacon urged as the "advancement" of knowledge. What he did in that role, as revealed in his message from the stars, is my reason for placing him at the start of this book. He did it with *system*.

Shutting Scholasticism's toolbox meant reaching for something to pick up the slack—something that could do double duty: both naming what was seen in the physical world *and* turning it into a message. In Galileo's hands, system fit the bill: the Jovian lunar "system" became evidence for a new "System" of the world. As it doubled itself to perform both roles, system turned Aristotelian logic inside out, from a rote form of "logical thinking" into a method for thinking logically about the "empirical world." Deployed in this manner, system made a new kind of knowledge while playing an instrumental role in forming our own epistemological practices. Today, we no longer even notice the doubling that allows us to speak of the telephone "system" as something both conceptual—a way to organize communications—and concrete—a physical wired or wireless network that can go "down."

Galileo's recounting of Copernicus's effort to find a new "way through" (the etymological meaning of *method*) can help us to recover the history of these habits. His account is part of his own effort to find a way forward from his observations in 1610 to the *Systems* manifesto two decades later. In 1615, a leading theologian and member of the Inquisition, Cardinal Robert Bellarmine, wrote a critique of *The Earth's Motion and Sun's Stability*, a book sent to him by its author in an effort to preempt a formal inquiry by the Inquisition. In a letter directed to Galileo as well as to the author, a friar named Paolo Anonio Foscarini, Bellarmine offered not a theological but an empirical, commonsense defense of heliocentrism:[5]

> When someone moves away from the shore, although it appears to him that the shore is moving away from him, nevertheless he knows that this is an error and corrects it, seeing clearly that the ship moves and not the shore; but in regard to the sun and the earth, no scientist has any need to correct the error, since he clearly experiences that the earth stands still and that the eye is not in error when it judges that the sun moves, as it also is not in error when it judges that the moon and the stars move. *And this is enough for now.* (Galilei [1615] 2008, 148, emphasis mine)

We have Galileo's response even though it was never published. Rather than arguing directly with the cardinal, he mediated their disagreement through a discussion of Copernicus.

Galileo's targets were those who claimed that Copernicus himself had been in the commonsense camp. In this interpretation, Copernicus floated the idea of the "earth's motion and the sun's stability" only as an "astronomer" needing a "hypothesis" that could save "appearances" and calculate "motions." But "he did not," they claimed, "believe it to be true in reality and nature." In fact, Galileo argued, the reverse was true—and that reversal turned on "system." In performing the "astronomer's task," Copernicus did not, asserted Galileo, find Ptolemy "lacking" but "completely satisfactory regarding the calculations and the appearances of the motions of all planets." Only after "taking off the clothes of a pure astronomer and putting on those of a contemplator of nature" did he "examine" whether that system "could also truly happen in the world and in nature. He found that in no way could such an arrangement of parts exist: although each by itself was well proportioned, when they were put together, the result was a monstrous chimera. And so he began to investigate what the system of the world could really be in nature" (Galilei 2008 [1615], 152).

Ventriloquizing through Copernicus, Galileo did not dispute Bellarmine on the basis of calculations or how motions empirically appeared. He simply asserted that, after Copernicus, the new "way through" was to work through systems in a new way. Contemplating the arrangement of parts into nonmonstrous wholes became the mind's method for ascertaining what "could really be."

As we have seen, the plural—system*s*—is important here. Systems came to mediate "the true and physical constitution of the world" as more physical systems were discovered requiring more contemplation of—and "investigat[ion]" into—the "nature" of system itself. This is the deep irony in Bellarmine wrapping up his case by asserting "that the moon and the stars move. And this is enough for now." That might, in fact, have seemed like enough when there appeared to be only a single moon in the sky, for then the message from the stars could be that it circled the earth like they did. But that argument had just passed its expiration date. Five years earlier, the message had changed: more moons meant that one system was not enough. "Systema," as Walter Ong (1956, 232) has observed, was

> an ancient Greek term, translatable perhaps as "set up" or organized, composite whole. ... Although this term had always been applicable to the Aristotelian or Ptolemaic cosmos, it was not particularly exploited in this connection for the reason that this cosmos was conceived as one unique system without even an imaginable competitor. The notion of wholeness was so inevitable that it was not particularly attended to. Hence the notion of system, an organized whole, was a rather uninteresting one.

System only became "interesting" when it became plural—initially when Copernicus pitted his new System against the established one. It became, to use Galileo's word, "really" interesting when systems started showing up inside of each other. Once they did, systems began to appear everywhere, not just in space. "Given its new currency," wrote Ong, system "took hold in connection with the universe of knowledge quite as quickly as it did in connection with the physical universe. ... By the end of the seventeenth century, the habit of thinking of philosophy itself ... in terms of a "system" had become well established" (Ong 1956, 235–236).

That habit was the condition of possibility for system's linear takeoff during the century that followed. Galileo had helped to set the stage by improving and using two kinds of tools: the spyglass as his physical instrument and method as his intellectual one. Each observation was recorded in both words and pictures in *Sidereus Nuncius*. Every page featured multiple pictures framed within lines of text (figure 1.2). The nascent

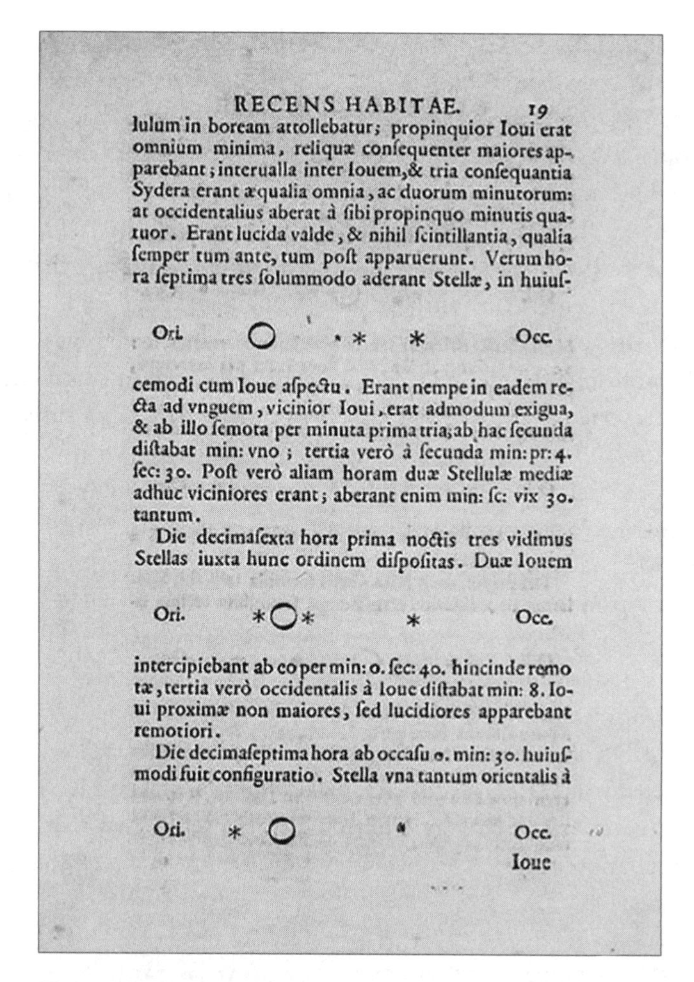
RECENS HABITAE. 19

Iulum in boream attollebatur; propinquior Ioui erat omnium minima, reliquæ consequenter maiores apparebant; interualla inter Iouem, & tria consequantia Sydera erant æqualia omnia, ac duorum minutorum: at occidentalius aberat à sibi propinquo minutis quatuor. Erant lucida valde, & nihil scintillantia, qualia semper tum ante, tum post apparuerunt. Verum hora septima tres solummodo aderant Stellæ, in huiuscemodi cum Ioue aspectu.

Erant nempe in eadem recta ad vnguem, vicinior Ioui, erat admodum exigua, & ab illo semota per minuta prima tria; ab hac secunda distabat min: vno; tertia verò à secunda min: pr: 4. sec: 30. Post verò aliam horam duæ Stellulæ mediæ adhuc viciniores erant; aberant enim min: sc: vix 30. tantum.

Die decimasexta hora prima noctis tres vidimus Stellas iuxta hunc ordinem dispositas. Duæ Iouem

intercipiebant ab eo per min: o. sec: 40. hincinde remotæ, tertia verò occidentalis à Ioue distabat min: 8. Ioui proximæ non maiores, sed lucidiores apparebant remotiori.

Die decimaseptima hora ab occasu o. min: 30. huiusmodi fuit configuratio. Stella vna tantum orientalis à Ioue

Figure 1.2
Sidereus Nuncius.

modernity of this method emerges if we imagine flipping one page after another, accelerating the sequence of frames. To treat this book as a kineograph is to receive its message as a visual effect—an effect that no human could actually "see": moons revolving around Jupiter. This "movement" effectively extrapolates a trend or pattern from Galileo's "data." And since this particular pattern is circular, it allows us to both fill in and predict additional data points.

Compare this message making to figure 1.1 at the start of this chapter. The "lately devised" tool at work there is an electronic database—a tool that extended my senses further than earlier iterations, such as isolated library collections and the *English Short Title Catalogue* (see Siskin 1998b, note 6). The observations it enabled have, like Galileo's, become visual data points in a temporal sequence—year after year rather than night after night and plotted by algorithmic systems rather than by hand. Extrapolating a pattern from those points has entailed, as it did in 1610, a visual effect, but in this case, the points have been "flipped" through the statistical method of regression, with the yield being a very straight line.

As in the earlier instance of pattern recognition, however, this "mechanical" reframing is only the first step in the new organon's translation of experience into knowledge. To prepare a message suitable for "inspection and contemplation" by others, Galileo then had to find his "way" from circles to systems. My task is to methodize my way from this single line into a history of system, and that entails both clarifying what "system" is and figuring out what *kind* of history to put it into. I have tried to prepare for the first endeavor by using Galileo to navigate to system's emergence, tracking the ways it first began to weave (and continues today to weave) in and out of being a thing in the world *and* being a way of constituting that world as a thing. I turn now to how—how system took hold, becoming powerful in both number and effect. This is a question of what makes things and what makes things happen. It is a question, in other words, of technology—both the technologies that have enabled the proliferation of systems and system's own status as a technology that proliferates.

TAKING HOLD—SYSTEM AS HANDSHAKE

> What it all came to, or comes to—how far what we call philosophy or theology or history or any other kind of knowledge can be related even analogously to a twirling set of bodies free-wheeling in space—is a matter which no one ever explained. The concept of system simply took unquestioned hold of the mind, applying itself everywhere. By the early eighteenth century, there is a real epidemic of systems.
> —Walter Ong, 1956[6]

To ask, as Walter Ong did, how system first took hold at the turn into the seventeenth century is to inquire into the number of things called

Figure 1.3
1640 frontispiece to Bacon's *Advancement of Learning.*

"system" at that time. That's not what we routinely do with system. Paying little attention to the possibility that system has a history to document, we stick the label onto all kinds of things. This can, of course, be very useful when we want to use our current understandings of system to corral things or clarify differences. Scholars of medieval and Renaissance works on memory, for example, compare and contrast them as "systems," even though the texts themselves do not use the term.[7] We can and we have learned much from deploying system in this manner, but it has also reinforced the assumption that system as we know it has simply always been around.

But it hasn't. We can use it now to connect past to present—memory systems back then to memory systems now—not because system remained the same but because it changed. It took hold at a moment and in a way that sutured the old to the new by bridging the gap identified by Ong: systems began to appear in space and in knowledge, occupying, at the same time, the physical world and the mental world. This was the Baconian agenda for renewal: the new organon differed from the old to the extent that it connected these realms in a newly interactive manner. For Bacon, knowing became about making. "To know the cause of a thing," as James Morrison has written of Bacon, "is to know how to bring about that thing oneself: … *Knowledge is ultimately the re-making of what nature itself makes or has made*" (Morrison 1977, 591–592).

The first half of the seventeenth century saw the instituting of this new relationship of active correspondence between knowledge and nature. By 1640, it crowns the frontispiece of a posthumous, revised version of Bacon's *Advancement of Knowledge* (figure 1.3) in the guise of two worlds—astronomically flanked in space by the sun and the moon—shaking hands: the visible world and the intellectual world. The handshake represents both the ongoing program for advancement and its progress since the first edition in 1605 and the publication of the *Novum Organum* in 1620.[8] The frontispiece to the latter had depicted a boat on an open ocean framed by the Pillars of Hercules marking the limits of the old world, accompanied by the motto, "Many will travel and knowledge will be increased." Two decades later, as Kaoukji and Jardine (2010) have pointed out, the pillars have become monuments to progress already made, with "plinths including the works of Bacon" topped by "elongated pyramids representing the universities of Oxford and Cambridge."[9] The pyramids, in turn, point directly at the now linked worlds.

As it began to occupy both of those worlds, system became an important formal basis for this newly interactive mode of producing knowledge. The presence of systems encouraged connections between them, and making more connections invited more claims of system. Bacon's plan for *Advancement* took shape at precisely the moment—the first decades of the seventeenth century—that system's Copernican turn to the plural began to be matched by the initial proliferation of systems in knowledge. In just a dozen years starting in 1600, for example, Bartholomäus Keckermann produced systems of logic, Hebrew grammar, theology, logic, ethics, politics, rhetoric, metaphysics, mathematics, astronomy, geography, and physics. And in 1613, just after Galileo found systems within systems in space, Johann Heinrich Alsted introduced the same phenomenon into the world of knowledge by turning Keckermann's output into a *Systema Systematum*. Alstead became well known for also seeding that world with two other variations of system. In 1610, the very year that the number of known lunar systems doubled, he published works on memory that he actually did call "mnemonic systems," including *Systema Mnemonicum Duplex*. And just two years before Galileo's labors yielded his comprehensive *Dialogue* on world systems, Alsted's efforts to compile systems methodically culminated in his influential *Encyclopaedia* of 1630.

Decade by decade system began to populate the cosmographies of both space and mind, a phenomenon little noticed and, as Ong noted, not explained. What's obscured it has not been a low profile but the opposite. System has saturated our worlds so successfully that we take it for granted. Ong and his mentor, Marshall McLuhan, spotted it only when the backdrop began to change. As system took hold in a new medium, the electronic, its relationship to the previous medium of print became visible as a history waiting to be told. How they told it, however, helps to explain not only why the history of system matters; it also suggests why, after half a century, there is still so much more to tell.

For Ong, system was part of a very big story: the "greatest shift in the way of conceiving knowledge between the ancient and the modern world." Knowledge, he argued, relocated from "discourse and hearing and persons" to "observation and sight and objects." This ushered in a turn in thinking to "visual, not auditory, analogies," as in the "drawing" of conclusions from the solitary "observation" of things rather than the "hearing" of a master and the sociability of "dialogue" (Ong 1956, 224). The historical coordinates of this "epistemological visualism," according to Ong, were "the transit from Aristotelian to Copernican space," on one axis and "the exploitation of letterpress printing" on the other (232, 228). And it's precisely at that intersection that system entered his picture. As the primary vehicle for negotiating that transit, system took hold with the help of print—"the central strategic operation in the procedure of visualizing knowledge at this time."

Ong's (1956) argument for this major epistemological change turns on the notion of seeing and using space in new ways, and it's in those alterations that system and print meet. Galileo found systems in space, but those systems also bounded space, turning everything within the boundaries of the system into its content. Print, in Ong's rendition, worked in the same manner, silencing the sound of words by shutting them into space "more adroitly than ever before":

> With one set of punches, one could move over bits of softer metal and strike out whole boxfuls of matrices. Casting from one set of matrices, one could produce whole fonts of type. With one font of type, one could set up an indefinite number of lines and compose an indefinite amount of type for making up an indefinite number of printing forms. From one form, one could print an indefinite number of pages simply by moving the paper into contact with the type and pressing it. Space had become pregnant with meaning.

When books became products of print, Ong argued, their pages became containers for that meaning, and their titles gradually changed from addresses to the reader to "labels like the labels on boxes" (228–229).

Speculating that this change in the way of "dealing with knowledge could not but affect the notions of what knowledge itself was," Ong (1956) imagined a logic of containment taking hold: if "knowledge could be 'contained' in books, why not in the mind as well?" Although Ong did not reference it, this is Bacon's handshake—and as I have described it: the visible, physical world of the book connecting to the intellectual world through the lingua franca of proliferating systems:

> The mind now "contains" knowledge, especially in the compartments of the various arts and sciences, which in turn may "contain" one another, and which

> all "contain" words. Discourse contains sentences, sentences contain phrases, phrases words, and words themselves contain ideas. (229)

The handshake was a signal that Galileo's message had been received: our worlds consist of systems within systems.

CHANGING OVER TIME—SYSTEM AS A GENRE

Handshakes with messages have resurfaced in our own moment of technological change, and once again in the company of systems. A handshake for computer systems is an exchange of standardized signals that control the start and end of the transmission of messages. But that connection between past and present is only one way that the handshake and the message I have just described can help us put system into history. Each of them points to a way of conceiving of system that allows us to track system through time and explain what it does.

The message of systems within systems tells us that the shift to the plural—to more and more systems—was enabled by what became a characteristic of system itself: lunar systems within larger systems meant system was *scalable*. Its capacity to assume different sizes meant that it could both adapt to changing situations and contexts and alter them as well as itself. This property has become particularly important with the onset of computing systems where scalability signals the ability of a system to handle a growing amount of work. Engaging system as a scalable technology, as I shall venture in the next section, can help us understand its persistence over time and its patterns of proliferation.

The handshake of worlds tells us that as system took hold at the turn into the seventeenth century, it changed. It changed not in the sense of becoming a different thing, but in terms of its changing interrelations with other kinds of things, such as technologies (print), objects (books), and concepts (the world). These alterations transpire through and are registered by shifts in features, how those features function, and the effects they produce. One ongoing feature of system, for example, has been boundaries. For system as the singular *systema* of the Ptolemaic cosmos, this feature functioned to convey inclusiveness—that everything was part of one thing. But for the systems of Bacon's handshake and of print, that same feature conveyed something different. It became a key component of Ong's logic of containment: boundedness turned knowledge into content. Highlighting boundaries also had other historical effects: it laid the groundwork, for example, for future salient distinctions, such as our contemporary concern with open versus closed systems.

Our term for something that takes form in relationship to a technology, and is identified by its characteristic features, as well as its interrelations with other forms, is *genre*. That term first appeared in English in the last two decades of the eighteenth century as a way of coping with the sudden and steep takeoff in print at that time (Siskin 2005, 818–822). As the language of "species" took hold in natural philosophy,[10] *genre* rose as the preferred term for discriminating kinds within the other realms of arts and letters. "With regard to the *genre*," reads the *OED*'s first citation from 1770, "I am of opinion that an English audience will not relish it so well as a more characteristic kind of comedy."[11]

Statements like these—and all other evidence of classification—are the basis of genre as I am invoking it here: not as a set of fixed, essentialistic definitions but as groupings that are made and remade in time. These historical groupings are, in Ralph Cohen's wonderfully efficient formulation, "empirical not logical." Arguing against the assumption that all members of a genre must share certain persistent, identifying markers, Cohen posits genres as purposeful and flexible. They "arise, change, and decline for historical reasons. And since each genre is composed of texts that accrue, the grouping is a process, not a determinate category." Genres are thus "open categories" that always form in relation to each other, constituting, at any particular historical moment, an interrelated community (Cohen 1986, 204, 207).

To engage system as a genre in this historical sense is a defining strategy of this book. Even in our best efforts to connect system to history, system has been strangely sublimated into an intellectual issue: an *idea* that carries and accumulates meanings rather than as an object that works in the world—or doesn't—to produce them. What has been missing is the turn to technological change: to make an object, you need a technology. Since, in our electronic age, systems materialize daily as networks and many other things, shouldn't we expect to find parallel sorts of embodiments in the past? And, if so, why haven't we attended to them?

Ong is the obvious and extraordinary exception here: his linking of system to print is still the clearest yes we have to the first question. But it also points to at least one possible answer to why we haven't continued to pay attention. The issue here is not a fault in the argument but a matter of trajectory. Ong did make an effort to anticipate what others might read into the link: "The use of printing need not be regarded as the *cause* of this shift of the focus of knowledge toward spatial analogies, but rather as a spectacular symptom of the general reorientation going on" (Ong 1956, 230, emphasis mine). "Need not," of course, does not mean that print couldn't be understood as causal, and the surrounding metaphors, such as an "epidemic" of systems, appear to invite that construal by mixing it with blame.

Marshall McLuhan explicitly accepted that invitation six years later by launching one of his deliberately provocative "probes":[12]

> The pseudo-dichotomies and visual quantities imposed on our psychology by print began in the seventeenth century to assume the character of consumer packages or "systems" of philosophy. They are of the kind that can be described and presented in a few minutes, but, thanks to the mesmerism of print, were to occupy the attention of generations. Philosophies from Descartes onward are diverse in the way in which a steam engine differs from a gas or diesel engine. (McLuhan 1962, 246)

As a genre, a McLuhan probe is not about truth but about consequences—about not being able to see things in quite the same way after the probe hits its target. This probe, for example, was not aimed at resolving specific issues in those "philosophies," but was intended instead to inspire second thoughts among those who have spent more than "a few minutes" attending to John Locke, Bishop Berkeley, or David Hume. McLuhan designed it to haunt those who study "ideas" with the possibility that they are missing something when they prioritize content over form.

The risk of any probe—however provocative—is that it turns out to be a one-off, and that has been the fate of this one. Any attention it might have brought to system was likely lost in the critique of technodeterminism that has preoccupied McLuhan's critics. His causal certainty that print "imposed" on us did bring Ong's story of system "onward" into the eighteenth century but as a tale of repetition that did not invite elaboration: the deus ex machina of the "engine(s)" draws the curtain on that story. Finding system as a formal pattern in eighteenth-century philosophy is a very good payoff for a probe. But what if we want to look further—to track, for example, how patterns change over time? Can new patterns emerge from old patterns? Why, for example, does system not only persist but proliferate as we snap out of—as McLuhan understands it—the mesmerism of print?

To give system a history that tracks it back to Aristotle and into space, through its changing interrelations in print, and then into the algorithmic and electronic requires a strategy that neither Ong nor McLuhan needed for their immediate purposes. I engage system as a genre to suit this broader agenda. Not only can it do the work of both Ong's most common descriptor, the "notion of system" and McLuhan's reference to system as "of the kind"; genre also provides a way to analyze patterns in relationship to each other and to their constitutive technologies.

Genre's origin in the late eighteenth-century takeoff of print in Britain[13] can also help us to explain our predilection for seeing system as an idea. The proliferation of texts was matched by an increase in the forms they assumed: the genres of written English multiplied in kind and in number. Some of those genres attracted considerable and ongoing attention, particularly those that became objects for specialized study within new disciplines forming at that time, as the novel did from the nascent profession of literary studies. Many others, however, slipped from view, their status as *written* objects obscured by what Raymond Williams called the "naturalization" of print—the historical process by which socially specific uses of a technology came to be seen as simply normal human activities.

The products of those activities also tended to lose their technological edge—their identities as embodied objects—some because they were not widespread and some because they were, like system, spread everywhere and particularly visible nowhere. Thus "system" comes to us from the eighteenth century as a familiar idea but not as a genre. Yet in the 122 times that *system* and its variants appear in the first edition of Johnson's *Dictionary* (1755), that is one of its most telling guises. The second quotation under the term *systematical*, for example, explains system by setting up what is to us a surprising but historically compelling contrast. "Now we deal much in essays," wrote Isaac Watts in *The Improvement of the Mind* (1741), "and unreasonably despise *systematical* learning; whereas our fathers had a just value for regularity and systems."

"Systems" versus "essays": this opposition turns, as suggested by Watts's subtitle—*A supplement to the Art of logick: containing a variety of remarks and rules for the attainment and communication of useful knowledge, in religion, in the sciences, and in common life*—on how best to produce, circulate, and consume knowledge. His implication, seized on by Johnson in defining system as the "reduc[tion]" of "many things" into a "regular" and "uni[ted]" "combination" and "order, is that essays entail a less regular ordering or reduction of things than systems do. In another quotation, from Boyle (1661), that distinction is echoed in the form of a preference: "I treat of the usefulness of writing books of essay, in comparison of that of writing systematically."

If you were a writer in the eighteenth century, you had to choose—not between these ideas or notions but between these genres. And that choice was crucial, for it determined the kind of knowledge you would produce. An essay then was not today's polished pearl but an irregular attempt—sometimes adventurous and always unfinished—"a loose sally," in Johnson's words. System, however, sought—and assumed the possibility of—completion, reducing "many things" to "order." To choose one or another was not just to indicate a stylistic preference but to make a statement about what could and should be known, and how.

Engaging system as a genre recovers for us this moment of choice and its consequences for the shaping of knowledge during and since the Enlightenment. And in combination with our new electronic databases, we can see these shapes with a new kind of clarity. We can, that is, count how many self-proclaimed systems appear in print and how often system is referenced in other kinds of texts. A brief example of the difference this makes is the assumption in earlier work that there must have been a turn from system in the later 1700s, particularly in the 1790s when, grouped with "theory" and "method," "system" was linked to France—in opposition to English "common sense" and "empiricism." This lineup of terms is David Simpson's, whose *Romanticism, Nationalism, and the Question of Theory* (1993) is our best treatment of system in the eighteenth century. He argues that system thus became a weapon in the discursive wars of English nationalism, particularly the conservative assault on radical thinking after the French Revolution.

Treating system as an idea, Simpson (1993) can attend brilliantly to the twists and turns of this tale as changes of meaning, detailing how each side tried "to capture the vocabulary" (60) of the other. The results were often paradox: the French as excessively devoted both to "reason" and to "sensibility" (76) and English literature as an "orderly disorder" (134). But, asserts Simpson, the general thrust of the English "nationalist tradition" is clear. By the 1790s it was "firmly set against system and theory": "As in government, so in argument and experiment the English idiom was hostile to hypotheses, schemas, and prescriptive constitutions. So pervasive was this idiom that it affected even writers apparently on the left" (52). The "systematic" became, in "literary" terms, "unmarketable" (170).

Faced with such hostility, system—as an idea—should have been in eclipse by 1798. And if we supplement Simpson's analysis and also think of system as a genre, the competition with essay might add to the logical expectation that systems and system writing should have been on the run. But the turn to genre also provides another way to ascertain what did happen, and the actual counts belie these expectations in startling ways. Not only did references to "system" maintain their steeply linear growth during the 1790s (figure 1.1); the percentage of texts with variants of *system* on their title pages[14] suddenly took an exponential turn at precisely that time (see figure 1.4). And if we limit the title page searches to just *system* and *essay* in the singular—and thus limit the effect of a striking late-century upsurge in collections of essay*s* (see appendix B)—*system* outpaces *essay* for the first time in the century in 1796, with the gap widening quickly to a doubling by century's end (see figure 1.5).

These results call for explanations that reveal system at work in ways that have simply not been visible in earlier studies—and that may run contrary to what system as an idea appears to tell us. As a genre, for example, system's prominence is a function not only of stand-alone systems but of another indicator of generic influence: how often the genre, in whole or in part, is embedded in other genres. The power of satire in the early eighteenth century, for example, is indexed by both the number of satires published and by the incorporation of satiric features in other forms—thus we classify Alexander Pope's *An Epistle to Dr. Arbuthnot* (1734) as a satiric epistle. To the question, "Where did systems go under the political pressures of the 1790s?" we might therefore answer, "The most vulnerable took cover within other genres." Such was the case with Britain's most

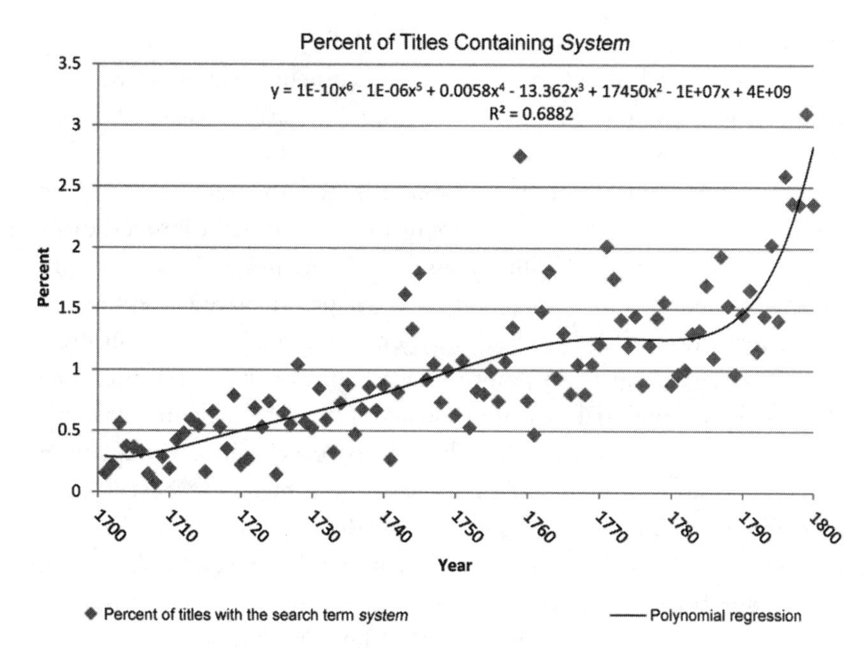

Figure 1.4
Using *Eighteenth Century Collections Online.*

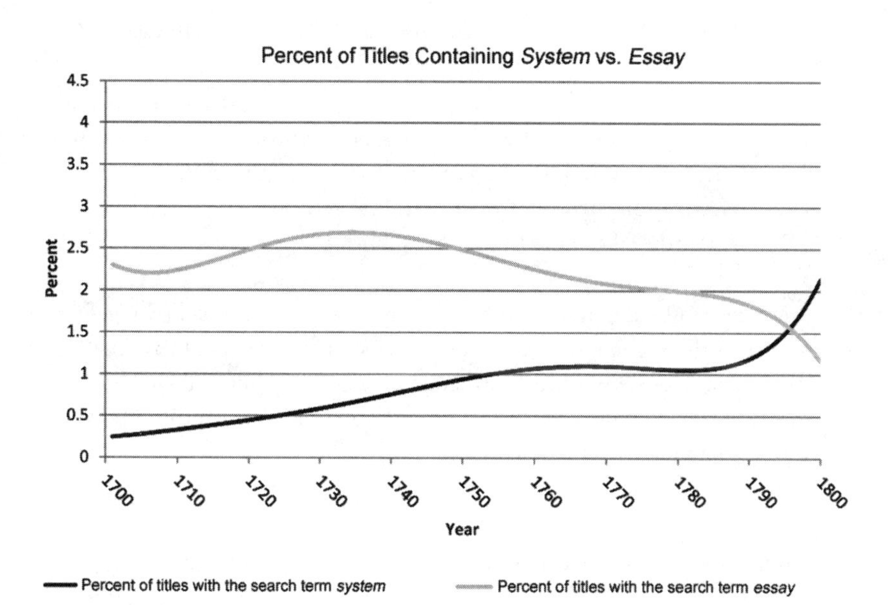

Figure 1.5
Using *Eighteenth Century Collections Online.*

famous political philosopher of that time, William Godwin. After seeing friend after radical friend jailed by the government, he chose to hide his next political system *in* a novel: *The Adventures of Caleb Williams Or Things as They Are* (1794).

Thomas Malthus, Godwin's contemporary, also strategically mixed genres, highlighting his system of population by putting it into a shorter, more accessible form in 1798: *An Essay on the Principle of Population*. This practice of generic embedding, I argue in chapter 4, played a central role in the formation of the modern disciplines as the Enlightenment drew to a close. Our counts also point to another factor that accelerated that process: the exponential increase in stand-alone, self-proclaimed systems. Instead of supporting encyclopedic efforts to contain and explain as much as possible, these new stand-alones were vehicles for specialization and professionalization, such as systems of education and the income tax. System, we find, participated so robustly in the shaping and reshaping of knowledge because one of its most persistent features—a feature that first came to the fore with the moons of Jupiter—has been its capacity to scale.

SHAPING KNOWLEDGE—SYSTEM AS A SCALABLE TECHNOLOGY

> I fancy mankind may … grow weary of preparation, and connection, and illustration, and all those arts by which a big book is made. If a man is to wait till he weaves anecdotes into a system, we may be long in getting them, and get but a few, in comparison of what we might get.
> —Samuel Johnson, 1773 (quoted by Boswell)[15]

> The limitations of physical books … brought us to think of long-form thought as the highest and most natural shape knowledge could assume. But what shape does networked knowledge tend toward? Short-form thought? Narrow-form thought? Or, perhaps the idea of shape gets in the way when we're trying to understand knowledge.
> —David Weinberger, *Too Big to Know* (2011)

The premise of this long-form big book[16] is that considering the shape of knowledge can help us to understand it.[17] What does get in the way is the idea that knowledge should have one shape—or that any particular shape is, intrinsically and ahistorically, a better shape for knowledge than any other. To make headway in understanding knowledge, we need to explain why a particular shape works at a particular moment and how what works changes over time.

From Galileo to Johnson to Weinberger, systems have been regularly made, used, and invoked in such explanations—and always in relation to their capacity to scale. Galileo's systems within systems become Johnson's systems made of anecdotes become, in Weinberger's requiem for books, an economic system whose causal capaciousness is asserted by the feature of system we explored earlier:

> Books … express the nature of knowledge committed to paper cut into pages without regard for the edges of ideas, bound together, printed in mass quantities, and distributed, all within boundaries set by an economic system. (Weinberger 2011, 96, 99–100)

Boundaries operate here on yet another scale from what we already noted in regard to the totality of *systema* and Ong's more localized containers of content. In this use of system, they are scaled up to provide an explanatory frame for the physical and historical regime of book knowledge.

In addition to boundaries, another persistent feature of those things that have been historically identified and grouped as systems is the relationship of parts and wholes. The *OED* etymology for *system* is "syn-prefix"—meaning "together," as in *synthesis*—plus "στα-, root of ἱστάναι to set up (see stand v.)." System is in this sense, then, something with parts that stand together as a whole. On this semantic platform, acts of organization transpire though establishing scale—what fits within what—with the potential of multiple scales: wholes becoming parts of larger wholes.

Scaling has been one of system's primary ongoing jobs—one of the characteristic ways that it works. I call system a scalable "technology" to emphasize that it has been used to perform that labor. In fact, these two terms—*system* and *technology*—are linked both chronologically and semantically. *Technology* made its first appearance in English in 1612, just two years after the debut of *system* cited earlier. Ong has tracked its journey through Greek and Latin, from Cicero using it to describe his treatment of grammar to Peter Ramus extending "technologia" to other arts in the late sixteenth century.

At that time, terms that we now associate with the sciences, such as *invention* as well as *technology*, were firmly ensconced in the literary domains of grammar and rhetoric. But in the late sixteenth century, under growing pressure to mount and execute a curriculum for Europe's expanding universities, Ramus played a central role in relocating them within a new classification of knowledge. For Ramus, as Ong emphasized, "the chief business of all classification is the classification of the arts and sciences themselves." Such classification is both the starting point of all philosophy and the means of teaching it at a large scale (Ong 2004, 197).

Seeding this new classification with technology enabled Ramus to play a key role in furthering Ong's "greatest shift": the turn, in the face of letterpress printing, from knowledge as a social product of sound and conversation to that which occupies the silent space of a page, the product of an internalized, self-referential "reason." *Technology* fit that agenda as a term that even in its etymology embodied that self-referentiality. It is the "speaking of" (-ology) a "techne," that is, an "art," or "method" or "system." And "system," as the form in which all parts must relate to the whole, became valued as a model for maintaining an internal dialogue: in Kevin Kelly's disarmingly simple description, "a system is anything that talks to itself" (Kelly 1994, 125). Technology, as the speaking of something that speaks to itself, thus became an embodiment of a silent and self-sufficient reason.

Where did the conversations that fueled old knowledge go? They ended up inside system—which, as a technology, first proliferated in the hands of Ramus's followers, such as Johann Alsted, the systematizer of Keckermann's systems and a pioneer of the encyclopedia. Ramists found systems and technologies of all kinds to be particularly useful in their curricular efforts, increasingly drawing them and "method" away from rhetoric or literary composition to the "logic" of "dialectic" as the "center and core of all philosophy" (Ong 2004, 197, 307, 145). As that logic materialized in print in the form of elaborate tree diagrams—diagrams that became Ramism's trademark—dialectic became increasingly associated with the quantitative: logic, that is, was no longer to be enacted in dialogue but in the measuring out of the world on the page.[18] Ramus, argued Timothy Reiss, had two goals:

> One was to programme old knowledge so as to prevent missteps in its learning. His spatialized grammar of dichotomies (aka Method) gave the needed machine. Another was to compute new knowledge in ways that, if by definition not absolutely certain, could still have the surety of an invariably ordered computing machine. Ramus and others increasingly hoped to achieve this goal through the mathematics of the quadrivium. (Reiss in Rhodes and Sawday 2000, 46)

Just a century after Ramus's murder in 1572, the dichotomies and tree diagrams had branched into the infinitely minute scale of the calculus, even as the "world" was scaled up into a universal "system" in the pages of Newton's *Principia*.

To engage system as a scalable technology is to get a sense of why that particular genre, among so many others, came to play such a central role in reshaping knowledge. By focusing on that specific feature of system, we can supplement Ong's speculations with concrete examples of why system was such a good fit for that job. "It was comforting," mused Ong,

> to think of oneself, or of one's enemy, as possessing a philosophical "system," something which whirled dazzlingly around a center in the mind like the Copernican spheres around the sun. … Such pictures could cover intellectual situations of which one knew really very little. The very looseness and inadequacy of the system metaphor was and is one of its greatest recommendations. (Ong 1956, 237)

This is a valuable insight. As we shall see, system does get deployed in "enemy" mode in the late eighteenth and early nineteenth centuries. It is, I argue, the secret history of Romanticism. But if "looseness and inadequacy" recommended system, then so too did its fit and flexibility as a scalable technology.

Bacon's generic formula for his instauration highlights the need for such a technology. Since he posited knowledge as starting anew from "things as they are," Bacon insisted that he had to start with what Weinberger (2011) would call short-form knowledge. Aphorisms and essays, he argued, were all that his method could legitimately manage in its early stages. Once the renewal gained momentum, however, it would have to scale up. To do that, required what he termed (in Latin) as a new "apparatus" (Bacon 1620, 24).[19] With natural history as his chosen form, Bacon's mode of scaling turned out to be quite literal: he added one history to another. The 1620 *Novum Organum* included a section titled "Preparation for a Natural and Experimental History," one "adequate to serve as a basis and foundation of True Philosophy." As I noted earlier, it consists, after an introduction featuring 10 "Aphorisms On Compiling a Primary History," of a "Catalogue" of 130 histories, featuring histories within histories—15 of which, for example, are "Histories of Species," while 88 are "Histories of Man" (Bacon [1620] 2000, 222–238).

Scaling up can be a daunting enterprise if the only apparatus is list making. Bacon's plan remained a plan, and he became famous for his arguments about method and power, not for the length of his lists. Unlike system, history was a genre that did not have features conducive to scaling up already on board. System, of course, had its own liabilities, depending on the job at hand. Bacon, for example, worried that system making was a particularly pernicious form of deduction: putting principles *before* things was an "anticipation of nature" that prevented the proper fixation of mind (Bacon [1620] 2000, 38). With that idea fixed in his own mind, he largely steered clear of system as a genre.

Although the systematical continued to draw suspicion and, as we shall see, continues even today to be an object of blame, system did become the genre of choice for scaling up into newly expanded "knowable spaces." I use that phrase to highlight how new resources—whether a better spyglass or print or a new method—extend not only what is known but, crucially, the range of what *can* be known. For natural philosophy that meant the extension of the "world" through the universal reach of natural laws such as the one governing gravity. Newton encoded the logic of that extension as a rule in a system. The second of the "Rules for the Study of Natural Philosophy" that head "The System of the World"—the Book that caps the *Principia*—was an aid for navigating the new reaches of knowable space:

> Rule 2 *The causes assigned to natural effects of the same kind must be, so far as possible, the same.*
>
> Examples are the cause of respiration in man and beast, or of the falling of stones in Europe and America, or of the light of a kitchen fire and the sun, or of the reflection of light on our earth and the planets. (Newton [1687] 1999, 795)

What was experienced as local now had purchase on the universal. If we understand light in the kitchen, we can understand lights in space. In

Newton's system, the laws of nature were as scalable as the genre in which they were communicated.

This confidence that the world could be known—even as there was more world to know—transformed the knowable space of moral philosophy as well. "Man" could now be the object of "science," argued Hume, and contemporaries recognized the same genre at work in both spaces. In the *Wealth of Nations*, claimed Thomas Pownall (1782), Adam Smith had constructed

> a system, that might fix some first principles in the most important of sciences, the knowledge of the human community, and its operations. That might become *principia* to the knowledge of politick operations; as Mathematicks are to Mechanicks, Astronomy, and the other Sciences. (Smith [1776] 1987, 337)[20]

From the start of his career, Smith's strategy had been to produce knowledge by exploiting system's scalable architecture. Part VII of *The Theory of Moral Sentiments* consists entirely of other "Systems." In a strategy repeated in his other major works, he folded those systems into his own—systems within systems—by "endeavouring to unfold" the "principles" from which "every system of morality that ever had any reputation in the world has, perhaps, ultimately been derived" (Smith [1759] 1976, 265).

Crucially, this enfolding was a temporal process that shaped knowledge over time by transforming wholes (previous systems) into parts (the new system). It thus enabled a mix of continuity and discontinuity, bringing existing ideas forward but altering their meanings and effects through this change of scale. Rescaling thus turned efforts at consolidation and preservation into vehicles for change. As we shall see, a major generic feature of the Enlightenment became the encyclopedic scaling up of separate systems into "complete" records of what was known.[21] Invariably, however, such efforts became the pretext for the publication of a new encyclopedia or yet another edition of the same one.

System thus participated in the process of error correction that the physicist David Deutsch (2011, 130) has identified as central to the Enlightenment effort to produce better explanations—explanations that benefit from adjustments to scale. Those adjustments made them harder to vary and thus better matched to the specific problems they were supposed to resolve. In helping to make those matches, system became a vehicle for converting information into knowledge per

Deutsch's very practical distinction: "Knowledge is information which, when it is physically embodied in a suitable environment, tends to cause itself to remain so."

Knowledge that is not suitably embodied, according to this understanding, does not remain knowledge; in Hank Williams's (2008) more aggressive formulation, "knowledge is information in non-disposable form." When the fit is suitable, the scalability of systems can shape knowledge in a very particular way, enabling in rare cases what Deutsch calls the "jump to universality": "the tendency of improving systems to undergo a sudden large increase in functionality, becoming universal in some domain." Deutsch's example is when the rules of a writing system are improved—as in our phonetic alphabet—so that the system becomes universal for that language—capable of representing every word in it (Deutsch 2011, 146). For an eighteenth-century example, consider Newton fitting the "world" into a "system" and writing "rules" that make that system capable of representing every movement within it.

That is not, of course, what happened or happens, to most systems. In 1785, Horace Walpole described systems as detonating rather than jumping:

> Philosophers make systems, and we simpletons collections and we are as wise as they—wiser perhaps for we know that in a few years our rarities will be dispersed at an auction; and they flatter themselves that their reveries will be immortal, which has happened to no system yet. A curiosity may rise in value; a system is exploded.[22]

My point here, and throughout this book, is not to tally jumps or explosions—to extol systems or to dismiss them—but to try to see them as they were and understand what they have done. And that entails engaging system in different ways—as a genre and as a scalable technology—and then identifying the histories in which its past can be told.

2 HISTORIES FOR SYSTEMS

> But so Matters fell out, and so I must relate them; and if any Reader is shocked at their appearing unnatural, I cannot help it. I must remind such Persons, that I am not writing a System, but a History, and I am not obliged to reconcile every Matter to the received Notions concerning Truth and Nature.
>
> —Henry Fielding, *Tom Jones*, 1749

SYSTEM AND HISTORY I

If this book engaged system as an idea, it could be labeled a history of system. I welcome that tag as a pointer for readers, but also as an opportunity to make two points crucial to the shape my argument will actually take:

- Although the history of ideas has been such a capacious and robust genre that we often simply think of it as "history" itself, it is but one kind of history. By that, I don't mean just the usual suspects now, such as social history or political history or the history of science, but of plurality in the past. Work on historiography, particularly in earlier periods, and especially Gianna Pomata and Nancy G. Siraisi's *Historia: Empiricism and Erudition in Early Modern Europe* (2005), has forcefully reminded us of the importance of understanding "history" as consisting historically of different kinds with different functions—history is, that is, a genre that, like all genres, is always subject to regrouping. And as we have seen with system, the resulting mixtures of features and functions change over time. Especially in moments of substantial change of resources, when reasons and opportunities for regrouping multiply—as in our current shift into the digital—writing a history requires more than choosing an object; we need to select with equal care the kinds of history into which that object will be written.

- When we think of kinds—of history as a genre as well as system as a genre—we need to keep Henry Fielding's remonstrance in mind. History and system have baggage—a past of changing interrelations that the preposition in "history of system" cannot overwrite.

Taken together, these issues are not impediments, but opportunities to use our historical ambitions to bring system closer. Using more than one kind of history can magnify the object by bringing those changing interrelations into focus. We can fashion a better spyglass.

In addition to improving an old tool, this chapter also offers a version of a new one. Its purpose is to help us explore and understand an extraordinary twist in the interrelations between system and history. As we use different kinds of histories to zoom in on system, what comes into focus is not just system but everything that comes into proximity with it—including other genres, such as history itself. The twist, that is, is a twist in the trajectories of how and how often those two genres were used during the eighteenth century. As indexed by the growing number of eighteenth-century title pages they share, system and history were on a collision course. My argument is that their intersection at the end of the century helps to explain a major event in the shaping of modern knowledge: the division of knowledge we call disciplinarity.

If we retrace the upward trajectory of system back to its Galilean launch, this twist becomes a bit less surprising. What we discover is that history became a genre of special interest at exactly the same time and for the same reason: to get knowledge going again. When, within a few years of Galileo's message, Francis Bacon described his plan for advancing knowledge, he did not claim, as we have seen, to be smarter or better than his predecessors. As with Galileo excusing Aristotle, ancients versus moderns, as I noted earlier, was not a contest for Bacon but history—a history that told of what he called his "good fortune" in living in a moment of different "resources" than they did. The resources that he cited were "mechanical things": printing, gunpowder, and the nautical compass. But Bacon recognized how easy it would be to squander those resources, so he warned against the idols of the mind and argued for the importance of method.

Most important, however, Bacon argued that no single mind—contra Descartes[1]—could advance knowledge on its own. The Great Instauration was collaborative to its core. Knowledge was stuck, Bacon argued, and we can only move it forward together. His major publications were more than monuments to himself and his own ideas; they were grant proposals: requests for funding for large-scale projects complete with appendixes detailing tasks and time lines. That's why the Royal Society considered him its founding father in a much more literal sense than we usually acknowledge: the members were not just honoring him as a great thinker but as the architect of the scientific institution as the *social* institution they now formed; their society was the embodiment of his vision of new knowledge as necessarily a joint undertaking.

Appealing to the powerful and the wealthy for support is not, of course, unusual—not back then or now. Bacon's particular request to King James, however, takes the form of a very specific and unusual proposal: it asks him to take "steps to ensure that a Natural and Experimental History be built up and completed" (Bacon [1620] 2000, 4–5). What is strange here is that Bacon does not want help building a building but building a genre—and that is what he asks for from the first page of his proposal to the last. In fact, the 1620 *Organum* ends with an extraordinary list of all the historical work that needs to be done—a list that bears the now telltale signs of a PI facing an imminent deadline:

> Now therefore we should move on to an outline account of *Particular Histories*.
>
> But as we are now distracted by business, we have only the time to append a Catalogue of titles of Particular Histories. As soon as we have the leisure for the task, we plan to give detailed instructions by putting the questions that most need to be investigated and written up in each history.

That catalogue begins with the "History of the Heavens; or Astronomy," veers earthward seven histories later to the "History of Clouds, as they are seen above," and continues to descend to number 12, the "History of all other things that fall or come down from above, and are generated above." By number 124 we reach the "History of Jugglers and Clowns" and, four places later, the "Miscellaneous History of Common Experiments which do not form a single Art" (Bacon [1620] 2000, 232–238).

Why history, and why so many of them? As we saw in my turn to Galileo, "system" became, to use Walter Ong's word, "interesting" only when it became plural—initially when Copernicus pitted his new system against the established one. It became, to use Galileo's own word, "really" interesting when systems started showing up inside each other. The same is true of history, as forcefully illustrated in Pomata and Siraisi's (2005) volume as they show history to be a genre that, like all genres, is a mixture of features and functions that change over time.

We can start plotting its interrelations with system by focusing on history's different relationship to the pedagogical and methodological priorities of Ramism and the new knowledge forwarded by Galileo and Bacon. Whereas system's "setting up" of parts and wholes could, as we have seen, structurally serve those priorities, history had what Donald R. Kelley calls a "detachment from form and structure" that did not lend itself to the new organon (Pomata and Siraisi 2005, 224). Galileo highlighted that issue by pointedly contrasting philosophers like himself to "historians" who relied on textual authority (225). That distance from form and structure did, however, afford history the flexibility to take on a variety of different tasks. In 1613—just three years after *system*'s debut in English and one year after *technology*'s—Rodolphus Goclenius's *Lexicon Philosophicum*, presented a fourfold definition of *history*. It could at that time be, to quote Kelley, a "simple description without demonstration," knowledge of particulars, a "syntagma" or specified "unit" of history, such as "the commemoration of antiquity," or observation from experience or the senses (212).

At the root of this multiplicity is a polarity of purpose with deep roots in the classical enterprise of *historia*. In Kelley's succinct telling,

> *historia* meant inquiry into specific but unspecified things and actions (*res, res gestai*), and soon the distinction between such things and the memories and reports thereof (*narratio rerum gestarum*) became confused, especially in English and the Romance languages. (213)

Pomata and Siraisi's collection is extraordinarily helpful in tracking how this distinction plays out in the genre's shifting mix of the empirical and the chronological. The volume as a whole demonstrates in detail that it was not until the late eighteenth century that temporality "moved" to the "core" of history. Many kinds of histories were simply *not* concerned with the passage of time. Here, for example, is *Encyclopaedia Britannica*'s definition of "history" in 1771:

> a description or recital of things as they are, *or* have been, in a continued orderly narration of the principal facts and circumstances thereof. (emphasis mine)

Notice that "have been"—the past tense—is only an option. The overall focus is on what we would now call the empirical: "things as they are."

To get a sense of how thoroughly this "or" shaped not only history but many other genres through the entire eighteenth century, compare William Godwin's title for his 1794 system-in-a-novel mentioned earlier to Delarivier Manley's title of 1714 in figure 2.1. "Adventures" or "History," Manley's binary, is flipped and recoded—per *Britannica*'s definition—into Godwin's "Things as They Are" or "Adventures." In both cases, *adventures* points to the retelling of the character's past, while *history* is the empirical "recital of things as they are." In using that phrase, Godwin, like *Britannica*, was echoing Francis Bacon who, as we have seen, foregrounded that phrase as a key to his plan to advance knowledge. For Bacon as well as for Godwin, "things as they are" invoked the empirical *as*

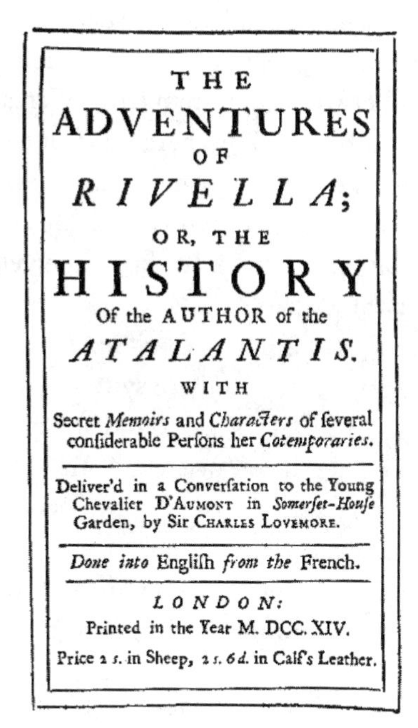

THE
ADVENTURES
OF
RIVELLA;
OR, THE
HISTORY
Of the AUTHOR of the
ATALANTIS.
WITH
Secret *Memoirs* and *Characters* of ſeveral conſiderable Perſons her *Cotemporaries*.

Deliver'd in a Converſation to the Young Chevalier D'AUMONT in *Somerſet-Houſe* Garden, by Sir CHARLES LOVEMORE.

Done into Engliſh *from the* French.

LONDON:
Printed in the Year M. DCC. XIV.
Price 2 *s.* in Sheep, 2 *s.* 6 *d.* in Calf's Leather.

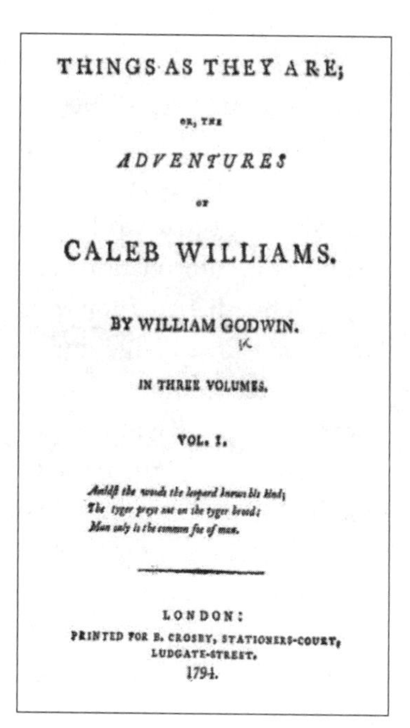

THINGS AS THEY ARE;
OR, THE
ADVENTURES
OF
CALEB WILLIAMS.

BY WILLIAM GODWIN.

IN THREE VOLUMES.

VOL. I.

Amidſt the woods the leopard knows his kind;
The tyger preys not on the tyger brood:
Man only is the common foe of man.

LONDON:
PRINTED FOR B. CROSBY, STATIONERS-COURT, LUDGATE-STREET.
1794.

Figure 2.1
Title pages from 1714 and 1794.

the historical. That's why Bacon claimed that his "advancement" into a future of new knowledge "must only be made by a natural *history*, and that of a new kind."

Although his mode of scaling up to that kind was the forbiddingly long list of histories that, from our perspective, strategically fell short, it's Bacon's motive that still speaks to us now: his insistence that history came in many kinds and that choosing mattered. For a history of system, this means that system is not the only variable. History, too, is a genre—a grouping that changes over time—and choosing how to engage it is as important to the outcome as choosing how to engage system. Choosing both is all the more important when we have two genres that have not only played important roles in shaping knowledge but have often done so in a competitive manner. It matters because—to put this as simply as possible—the competition hasn't ended. What we write about system now is shaped by how history and system have interrelated in the past and how they continue to do so in the present.

In most books, what others write about the same topic is either cordoned off in introductory comments or engaged at length, usually in the form of elaboration or argument. But since my focus on system is on how it shaped knowledge, the shapes that other work on system have taken are part of my inquiry. I bring this issue of other work up here—rather than in a preliminary meta-statement or as argument—because what illuminates the shapes of that work and the shape of my own is the subject of this chapter: the interrelations of history and system. In those interrelations, system has shaped knowledge about itself.

I have raised already, in comparing my findings to David Simpson's in chapter 1, the relationship of my enterprise to the study of system within the history of ideas. I will turn shortly to the story of how that type of history emerged at the turn into the nineteenth century to assume such a prominent place in the discipline of history. The rise of ideas as a primary currency of history went hand in hand, I shall argue, with history's changing relationship to system—specifically a transfer in agency from system to the modern subject. Three of the most prominent forms of system scholarship today also bear telling connections to the changing interrelations of that period. Immanuel Wallerstein's "modern world system" has been an effort to describe the rise of capitalism as the dividing of the world into geographical parts, with each part defined by the function it performs in a unified global economic system. Niklas Luhmann's work in systems theory has been an effort to remake Max Weber's rationalization narrative of modernity into a tale of functional differentiation through autopoetic systems. Cybernetics, most closely identified with Norbert Wiener, has also used system's feedback loops to explore issues of control arising out of the changing relationship between man and machine.

These have been valuable undertakings, all three of which abstract system from its status as a genre to put it to work as a conceptual tool explaining the conditions of modernity: the economic and political, in particular, for Wallerstein, the social and psychic, for Luhmann, and the technological for Wiener. Since these scholars do not do what I do—they use system to produce knowledge, while I study it—my business with them in this book is not critique but explaining how and why system ended up in their hands. To that end, I return system to its status as a genre so that we can study it through its interrelations with other genres, particularly—for our present purposes—its changing relationship to history.

THE TECTONIC MAPS

How do we identify and grasp that relationship? System and history, as we have just seen, were genres launched at the same time to address the same problem: how to pull out of a stall and scale up to the good fortune of new resources. But recovering what happened next—and the consequences—is no easy task, so it's time to use our new computational resources and locate system and history on our Tectonic maps. I call them "Tectonic" because the different shapes in each map resemble the plates that float on the surface of the earth.[2] These shapes, although irregular, fit together into what geologists would call a Pangea or supercontinent. The analogy is, of course, only a starting point for grasping what the maps represent and what they can do for us. Take a look at the first one (figure 2.2).

This map is the product of my collaboration with my research colleague Mark Algee-Hewitt, who in building it has drawn on work in information visualization at Bell Laboratories (see appendix A). Tectonics is in ongoing development—not only in the conventional sense of the technology getting better and the maps much prettier, but also in the

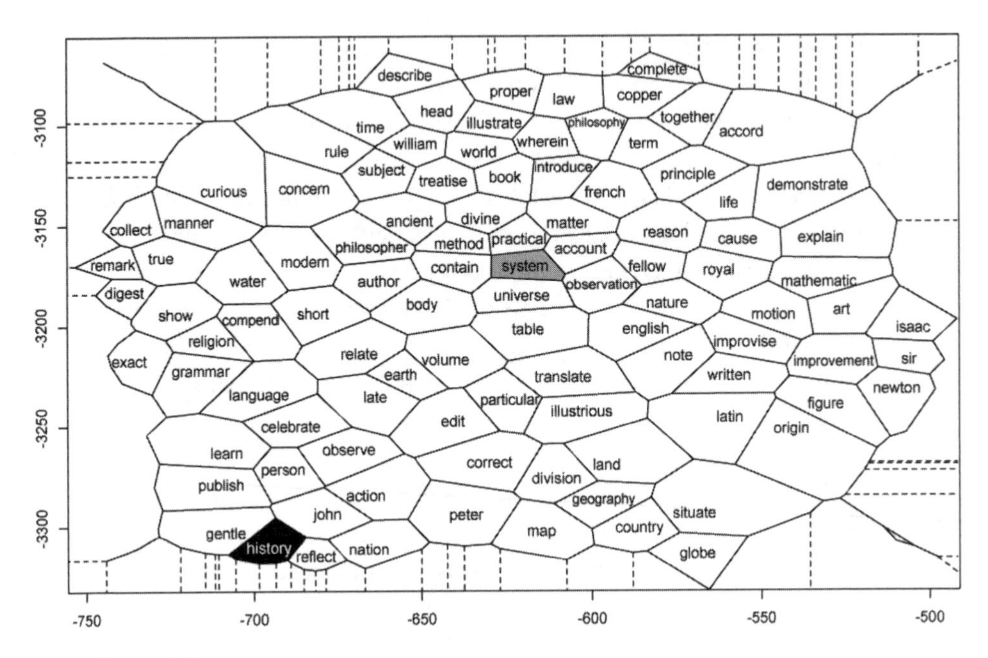

Figure 2.2
System in relationship to other terms with which it shares title pages, 1700–1739.

sense of how we develop in response. Like all new technologies, it will take time for it to take hold—for you and me and it to find its primary forms and functions. In a sense, it will tell us what to do next.[3]

Based on all of the title pages in *Eighteenth Century Collections Online* that contain variations of the word *system*, it tells us which other words also appear on those title pages. Each word receives a "distance" score from every other word, based on the likelihood or probability that both would appear within the same title in which *system* appears. These are stem searches, meaning that they catch all variants of the word—with one exception. Since our focus is on the relationship of *history* to *system*, we have given *history* special attention. Its variants occupy two plates: one, as with *system*, is the noun, both singular and plural; the other is for the modifiers, *historical* and *historically*.

Every plate is bounded by the words with which it most often shares title pages with *system*. The diagrams thus map both adjacencies *and* clusters—and the map itself has what we might call cardinality. Words in the northeast, as explained more fully in appendix A, are much more likely to appear with each other on system title pages than to any word in the southwest corner. Having *system*—as the common denominator—at the center of the map translates the frequency with which words share title pages with *system* into distance from the center. Every word's location is thus a function of that frequency *and* its clustering cardinality. As the date ranges change from map to map—moving sequentially through the eighteenth century—we can track which words and clusters in the descriptive title pages of that time[4] fall in and out of proximity with *system*.

In the first four decades of the eighteenth century, as mapped in figure 2.2, we find *history* at the periphery.[5] *System* is surrounded more closely by terms associated with Newtonian natural philosophy, including *nature, account, method, universe, observation, and matter, table,* and *body*. At a farther distance, even Newton's *nameplates* make it onto the map: "Sir" "Isaac" "Newton."

In the next third of the century, however, they fall off the edge (into the mouth not of a monster but of statistical insignificance), and a number of the Newtonian terms give way to a new set of systems that package "knowledge" into a "variety" of "introductory" "accounts" and "descriptions" that include "illustrations" and "copperplate" "engravings" (figure 2.3). This surge in "systems" that "illustrate" "knowledge" arises at the midcentury moment when the visual legacy of Ramism intersected with the scaling up both of system's encyclopedic ambitions and of the market for and technologies of print. As that confluence contributed to what John Bender and Michael Marrinan (2010) describe as the "culture of diagram," "history" was also on the move. In this slice of the eighteenth century, it halved its earlier distance from the center, approaching system in the company of a new set of terms: *history*, when sharing title pages with *system*, increasingly took on the task of "shew[ing]" things, particularly things that were "curious" or more "accurate."

What happened next, in the final two decades of the eighteenth century, is what I am calling the collision (figure 2.4).[6] Could this have been predicted—predicted in the manner that geologists are working hard to predict earthquakes today? What we can say is that *system* and *history* shared an initial push and direction from Galileo's and Bacon's contemporaneous efforts to advance knowledge—and that after the

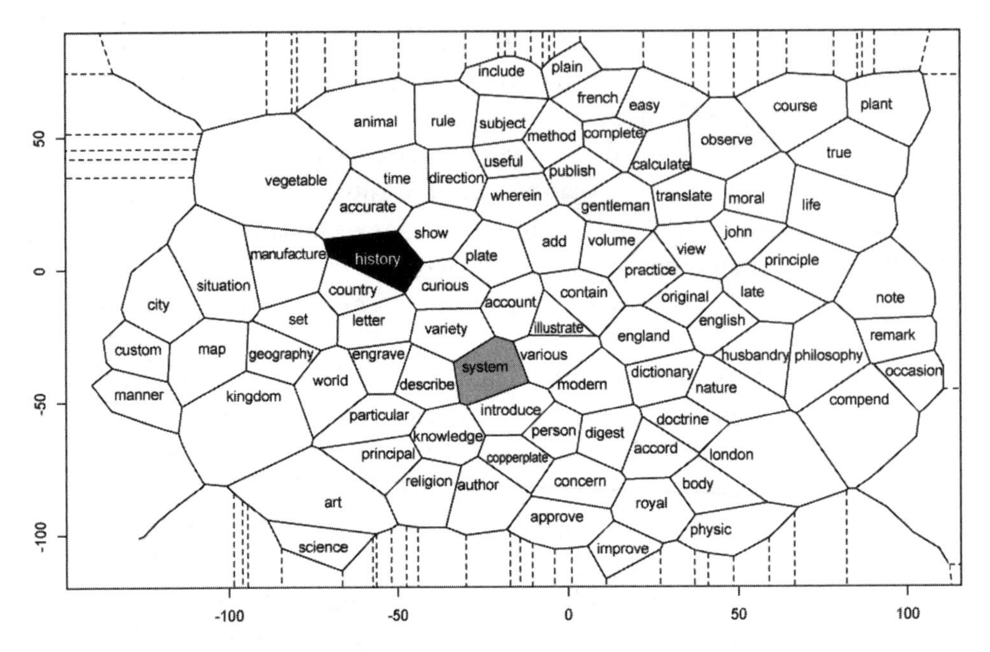

Figure 2.3
System in relationship to other terms with which it shares title pages, 1740–1779.

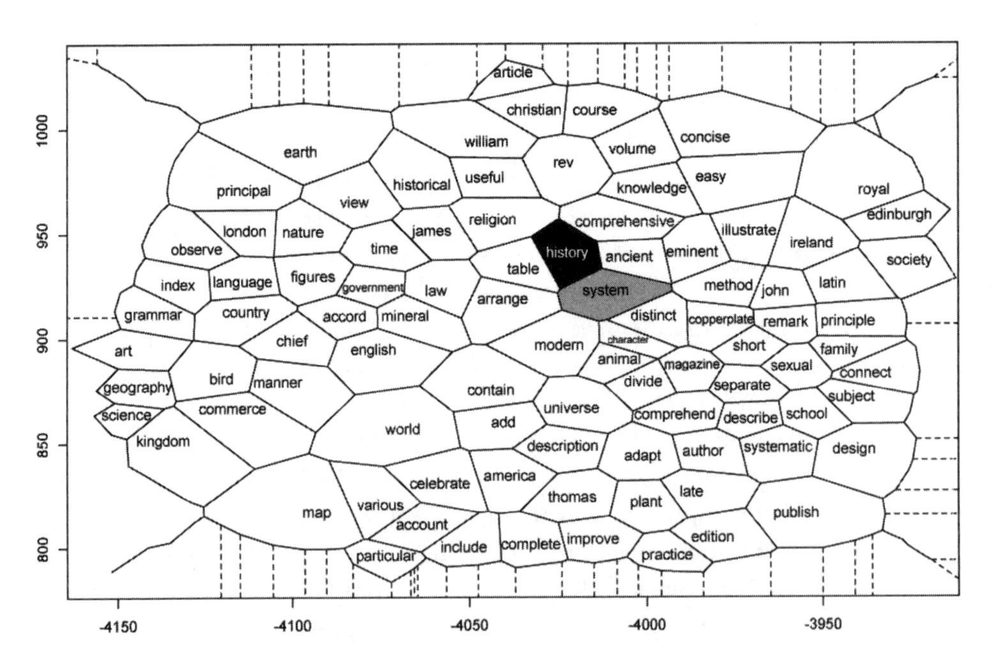

Figure 2.4
System in relationship to other terms with which it shares title pages, 1779–1800.

advancement we call Enlightenment, their trajectories intersected. What Tectonics does is materialize that intersection on a broad evidentiary base that—in its specificity and physicality (2,679 shared title pages)—can help us to understand better what happened and why. It offers, for example, an index to the force of that collision in the form of *historical** appearing on the map for the first time—an indication that the adjectival and adverbial constructions also now share more title pages with *system*, in what we might call the "generic modifier" format—as in "historical systems."[7]

By mapping adjacencies and clusters, Tectonics also identifies and visualizes the conditions of possibility for new forms of knowledge and understanding. In this case, the meeting, or collision, of system and history formed what I would term a platform—a generic platform—on which knowledge could and did divide into the modern disciplines. In terms of our understanding how the shape of knowledge changes, the stakes here are high. Citing evidence from William Clark (2006), Geoffrey Lloyd (2009), and Donald Kelley (1997), Simon Schaffer argues that "sorting out the geographical and chronological schemes of disciplinary power remains the analyst's task, since one cannot be sure exactly when there were disciplines to connect nor when they looked as they do now" (Barry and Born 2013, 2–3). As Schaffer points out, for example, "Robert Boyle systematically used the term *discipline*." But the use that Schaffer highlights illustrates the need to historicize rather than to assume the presence of our understanding of the term "discipline" back then.

When Boyle describes himself as "having invaded Anatomy, a Discipline which they ['Physitians'] challenge to themselves" (Boyle 1660, 382–383), his use of the term was more a matter of occupational gatekeeping among the professions than a statement about the nature of that knowledge. This was not an invoking of an overall organization of knowledge divided into narrow-but-deep kinds. What Tectonics offers is a way to historicize that reorganization by identifying the condition of possibility of those kinds. Their incarnation was enabled, I am arguing, by the historical formation of the platform visualized by Tectonics: the coming together of system and history.

Tectonics is, then, per Galileo, a new spyglass for zooming in—for charting spatially over time the shifting adjacencies that enable larger-scale change, including not only the reorganization of knowledge but also, per Peter de Bolla's work on the "division of labor" and "human rights," the formation of new concepts (Siskin and Warner 2010, 87–101; de Bolla 2013). Calling this method of materializing change "Tectonics" points as well to these shifting configurations as an act of *building*, the term's etymological root. A tectonic form is—as in an *OED* cite from 1903—"A form produced ... by the *exigencies* of construction."[8] Thus another payoff for Tectonics is that it can help us understand what is being built and under what conditions.

What were the exigencies that drove history into system and onto shared title pages? We focused the third map on just the last two decades of the eighteenth century because of two quantitative phenomena of that period of time: the overall takeoff in print and the increase in stand-alone, self-proclaimed systems as vehicles for specialization and professionalization (e.g., a system of the income tax, a system of education). In this growing, and thus more and more crowded, market for print, all kinds were under increasing pressure to lay claim to their part of it. Even the most comprehensive categories assumed more specialized forms.

Specialization, however, was not a simple matter of isolation but of renegotiating boundaries and functions with other forms. History's intersection with system was just such a negotiation, the terms of which we can find inscribed on the title pages identified by our maps. By scaling up to thousands of title pages, those maps also allow us to scale down—to zoom into specific examples of this new adjacency. We can see that under the pressure of those final decades, system and history, to use a word from one of the title pages in figure 2.5, entered into a "gratifying" relationship.

For Mrs. Dards's catalogue, on the one hand, the gratification comes from a "new system" interrelating with "history" by appealing to history's *empirical* mode of valuing "objects" "minutely described." On the other hand, Thomas Bankes's "NEW AND AUTHENTIC SYSTEM" of 1788 is enhanced by "Complete HISTORY" in its *chronological* mode of accounting for events that "have taken place." That temporal dimension provides system with the opportunity to exercise its scalability across time as well as space, for, with the past mixed in, Bankes's "SYSTEM" becomes a new "WHOLE" comprehending "EVERY THING WORTHY OF NOTICE THROUGHOUT THE WHOLE FACE OF NATURE, BOTH BY LAND AND WATER."

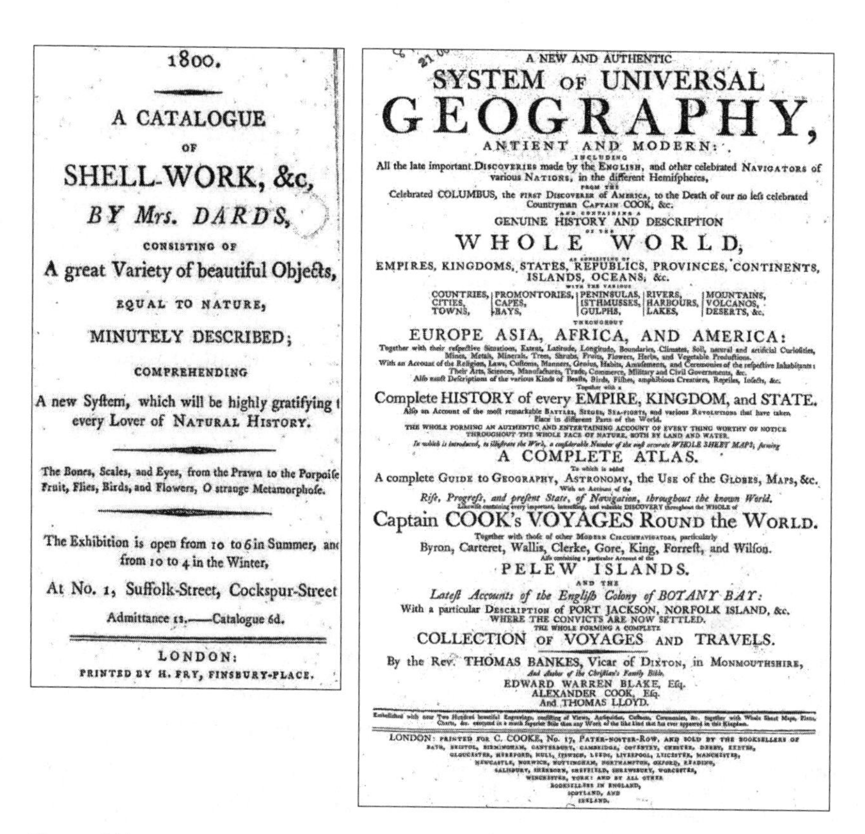

1800.

A CATALOGUE

OF

SHELL-WORK, &c,

BY Mrs. DARDS,

CONSISTING OF

A great Variety of beautiful Objeƒts,

EQUAL TO NATURE,

MINUTELY DESCRIBED;

COMPREHENDING

A new Syſtem, which will be highly gratifying t[o]
every Lover of NATURAL HISTORY.

The Bones, Scales, and Eyes, from the Prawn to the Porpoiſe
Fruit, Flies, Birds, and Flowers, O ſtrange Metamorphoſe.

The Exhibition is open from 10 to 6 in Summer, an[d]
from 10 to 4 in the Winter,

At No. 1, Suffolk-Street, Cockspur-Street

Admittance 1s.——Catalogue 6d.

LONDON:
PRINTED BY H. FRY, FINSBURY-PLACE.

A NEW AND AUTHENTIC
SYSTEM OF UNIVERSAL
GEOGRAPHY,
ANTIENT AND MODERN:
INCLUDING
All the late important DISCOVERIES made by the ENGLISH, and other celebrated NAVIGATORS of various NATIONS, in the different Hemiſpheres,
FROM THE
Celebrated COLUMBUS, the FIRST DISCOVERER of AMERICA, to the Death of our no leſs celebrated Countryman CAPTAIN COOK, &c.
AND CONTAINING A
GENUINE HISTORY AND DESCRIPTION
OF THE
WHOLE WORLD,
AS CONSISTING OF
EMPIRES, KINGDOMS, STATES, REPUBLICS, PROVINCES, CONTINENTS, ISLANDS, OCEANS, &c.
WITH THE VARIOUS
COUNTRIES, CITIES, TOWNS, | PROMONTORIES, CAPES, BAYS, | PENINSULAS, ISTHMUSSES, GULPHS, | RIVERS, HARBOURS, LAKES, | MOUNTAINS, VOLCANOS, DESERTS, &c.
THROUGHOUT
EUROPE ASIA, AFRICA, AND AMERICA:
Together with their reſpective Situations, Extent, Latitude, Longitude, Boundaries, Climates, Soil, natural and artificial Curioſities, Mines, Metals, Minerals, Trees, Shrubs, Fruits, Flowers, Herbs, and Vegetable Productions.
With an Account of the Religion, Laws, Cuſtoms, Manners, Genius, Habits, Amuſements, and Ceremonies of the reſpective Inhabitants: Their Arts, Sciences, Manufactures, Trade, Commerce, Military and Civil Governments, &c.
Alſo exact Deſcriptions of the various Kinds of Beaſts, Birds, Fiſhes, amphibious Creatures, Reptiles, Inſects, &c.
Together with a
Complete HISTORY of every EMPIRE, KINGDOM, and STATE.
Alſo an Account of the moſt remarkable BATTLES, SIEGES, SEA-FIGHTS, and various REVOLUTIONS that have taken Place in different Parts of the World.
THE WHOLE FORMING AN AUTHENTIC AND ENTERTAINING ACCOUNT OF EVERY THING WORTHY OF NOTICE THROUGHOUT THE WHOLE FACE OF NATURE, BOTH BY LAND AND WATER.
In which is introduced, to illuſtrate the Work, a conſiderable Number of the moſt accurate WHOLE SHEET MAPS; forming
A COMPLETE ATLAS.
To which is added
A complete GUIDE to GEOGRAPHY, ASTRONOMY, the USE of the GLOBES, MAPS, &c.
With an Account of the
Riſe, Progreſs, and preſent State, of Navigation, throughout the known World.
Likewiſe containing every important, intereſting, and valuable DISCOVERY throughout the WHOLE of
Captain COOK's VOYAGES ROUND the WORLD.
Together with thoſe of other MODERN CIRCUMNAVIGATORS, particularly
Byron, Carteret, Wallis, Clerke, Gore, King, Forreſt, and Wilſon.
Alſo containing a particular Account of the
PELEW ISLANDS.
AND THE
Lateſt Accounts of the Engliſh Colony of BOTANY BAY:
With a particular DESCRIPTION of PORT JACKSON, NORFOLK ISLAND, &c.
WHERE THE CONVICTS ARE NOW SETTLED.
THE WHOLE FORMING A COMPLETE
COLLECTION OF VOYAGES AND TRAVELS.
By the Rev. THOMAS BANKES, Vicar of DIXTON, in MONMOUTHSHIRE,
And Author of the Chriſtian's Family Bible,
EDWARD WARREN BLAKE, Eſq.
ALEXANDER COOK, Eſq.
And THOMAS LLOYD.
Embelliſhed with near Two Hundred beautiful Engravings, conſiſting of Views, Antiquities, Cuſtoms, Ceremonies, &c. together with Whole Sheet Maps, Plans, Charts, &c. executed in a much ſuperior Stile than any Work of the like kind that has ever appeared in this Kingdom.
LONDON: PRINTED FOR C. COOKE, No. 17, PATER-NOSTER-ROW, AND SOLD BY THE BOOKSELLERS OF BATH, BRISTOL, BIRMINGHAM, CANTERBURY, CAMBRIDGE, COVENTRY, CHESTER, DERBY, EXETER, GLOUCESTER, HEREFORD, HULL, IPSWICH, LEEDS, LIVERPOOL, LEICESTER, MANCHESTER, NEWCASTLE, NORWICH, NOTTINGHAM, NORTHAMPTON, OXFORD, READING, SALISBURY, SHERBORN, SHEFFIELD, SHREWSBURY, WORCESTER, WINCHESTER, YORK: AND BY ALL OTHER BOOKSELLERS IN ENGLAND, SCOTLAND, AND IRELAND.

Figure 2.5
Title pages containing *system* and *history.*

As history met system under the exigencies of those last two decades, the two genres rebuilt each other—and in the process, they reshaped other subject areas from branches of philosophy into stand-alone specialties. Both of these shared title pages announced such specialized systems—of biology and of geography—both of them becoming part of that genre's end-of-century surge and of knowledge's new shape. In Galilean terms, there were more systems within systems. In turn, system's scalability—its capacity to scale up to the "UNIVERSAL" for Bankes (1788) and to a "Variety" that was "EQUAL TO NATURE" for Mrs. Dards (1800)—opened new realms to the professing of history. Under the competitive exigencies that helped to specialize and professionalize system, more systematic histories took on more, with more becoming the mark of its changing status.

In Bankes's geography, history scaled up to the "whole" world, mounting new efforts at world and universal history that extended and rationalized earlier biblical and monarchical chronologies. A century after Newton had used a universal law of rising and falling to give a new coherence to the "system of the world" (1687), Edward Gibbon wrote a *History of the Decline and Fall* (1776–1788) and Alexander Tytler published a *Plan* for a *Universal History, Ancient and Modern* (1782).

With system now mediating the empirical connection to things—absolving history of that function—this was the moment in which temporality completed its move to history's core. New knowledge groupings now had empirical content arranged systematically *and* their own chronological narratives to define them and to differentiate them from each other. And they also now had professionals specializing in the writing and recitation of those narratives. Both Gibbon and Tytler were professors, and Tytler's *Plan* was for a "course of lectures."

To corroborate the phenomenon I have just described—the tectonic formation of the narrow-but-deep disciplines of modernity at the late eighteenth-century intersection of system and history—we need to be able to observe it from a different generic standpoint. The perspective that offers the greatest clarity is, not surprisingly, from one of the most prominent and proliferating genres of those final decades of the eighteenth century: the encyclopedia. When the first edition of *Encyclopaedia Britannica* appeared between 1768 and 1771, its editor explicitly denounced both organizing knowledge in alphabetical order and comprehensive diagrams as "repugnant" "folly." Materials from all of the sciences and arts were instead "digested" into substantial "Treatises or Systems" that proved to be *Britannica*'s most striking and consequential formal innovation: the consolidation of alphabetically organized material into an array of subject-specific "arrangements" that anticipated the modern curriculum. In effect, the encyclopedia became the home of protodisciplinary systems (*Encyclopaedia Britannica* 1771, v; *Encyclopaedia Britannica*, 1797, x).

As *Britannica* grew during the decades of the takeoff in print—almost quadrupling from 2,391 pages in the 1771 first edition to 8,595 in the 1784 second edition and then almost doubling again into the 14,579 pages of the third—"more" produced difference. Not only did the number of protodisciplines increase with the number of pages, but their boundaries thickened. What had been only digested was now defined and deepened. By the third edition, each of these "distinct" new systems was now packaged, proclaimed the title page, with its own "*History*, Theory, and Practice." This move sounds innocuous until Tectonics helps us to visualize what was at stake.

When finally packed together at the end of the century, system and history formed a new generic platform—a platform on which the encyclopedia's new knowledge groupings could institute their differences by telling their *own* tales. *Britannica* could now claim that it contained "Full EXPLANATIONS" of the "VARIOUS *DETACHED* PARTS OF KNOWLEDGE" (emphasis mine). What Galileo observed in the stars—that the system of the physical world was a world full of systems—the reader of *Britannica* could observe on the page: the system of the intellectual world was becoming, edition by edition, a world full of new systems that are now our disciplinary and departmental homes.

We may think or hope those homes are on solid ground, but the condition of possibility I have just described was a collision—and collisions in tectonics are always the first steps toward other collisions. Think of the late twentieth-century calls for interdisciplinarity as preliminary tremors—tremors that remind us that the core message of Tectonics (indeed, tectonics of any kind) is always instability. To say, as Bacon did when *system* was first entering English, that knowledge was "stuck" may have suggested that things were static—that nothing was happening. But in the context of tectonics, stuck signals change—it *is* the enabling condition of change—the state in which pressures build and forces multiply. Looking back from the late eighteenth-century collision I've just described, we can see Galileo's systems and Bacon's histories multiply until their collisions formed a new ground for knowledge. Standing on that ground, Tectonics gives us a new perspective on how we got there—and, in doing so, it asks us to anticipate a near future in which that ground will shift from under our feet.

HISTORY AMONG THE DISCIPLINES—THE NARROWING

After the collision, what Donald Kelley calls "the career of world or universal history" (Pomata and Siraisi 2005, 229) was set; its later twists and turns have all been efforts to build on the tectonic proximity of system and history. For Wallerstein (2004), for example, the world system was a product of history configured by capitalism. For Luhmann (1997), however, system was not a product but itself the active agent. "Under modern conditions," he argued,

> the global system is a society, in which all internal boundaries can be contested and all solidarities shift. All internal boundaries depend upon the self-organization of subsystems and no longer on an `origin' in history or on the nature or logic of the encompassing system. (72)

Here yet again is yet another version of what Galileo sighted through his spyglass: systems within systems. Luhmann's twist on the scalability of system was to insist that systems made themselves by using "their own output as input." History could not, of course, be the origin of autopoietic systems—for they are their own origin—but by no means did history drift off from system. In Luhmann, history (in its modern sense with time at its core) became the condition of possibility for autopoiesis itself as a recursive and thus necessarily temporal process.

Neither Wallerstein nor Luhmann represents the norm for writing history or for writing about system. For that we need to turn in the direction of the history of ideas. As I suggested earlier, the rise of ideas as a primary currency of history went hand-in-hand with the tale I have just told of history's changing relationship to system. The two are linked by transfers in agency between system and the modern subject. Luhmann's work can be our station for that transfer, since readers reaching for a way into his arguments have repeatedly grasped what looks like a straightforward substitution. In Linda C. Brigham's words echoing those of William Rasch and Cary Wolfe, "systems" in Luhmann "stands in the place of the old subject, and 'environments' replaces the old object" (Brigham 1996; Rasch and Wolfe 1995).

What enables this switch is that Luhmann is granting system autopoietic agency, and thus qualifying it to take over causation from the subject. System became a place to "send the mail," Brigham (1996) writes, once we recognized that "the address of the Enlightenment Subject has been vacant for a long time." But system and subject have actually been in a very long-term relationship, as I detail in chapter 7. Our focus here is on their interaction during the moment of history's modernization; the subject, as well as system, I will argue, played a role in that transformation.

To bring the subject into the picture, we need to pick up that process where we left off in the 1780s and then take it forward into the following decades. In 1784, while, Gibbon and Tytler were busy with their systematic scaling up of world history, William Jones wrote a "Discourse on the Institution" of the Asiatic Society that was later published in its *Transactions*. In discussing the kinds of knowledge he hoped the society would produce, Jones echoed Francis Bacon's classification of man's knowledge from 1605: "history to his memory, poesy to his imagination, and philosophy to his reason" (Bacon [1605] 2008, 175). This scaffolding for the new organon had persisted as a framework for new knowledge ventures thanks not only to the success of the Royal Society (the explicit model for this Asiatic one), but also Peter Shaw's new edition and translation of Bacon in 1733. Linking "memory, reason, and imagination" to "history, science and art," Jones, like Bankes in his *Geography*, set a new worldly agenda for history: it "comprehends either an account of natural productions, or the genuine records of empires and states" (Asiatic Society 1799, x, xiii).

For historians seeking to comprehend, Asia was the subject of daydreams. "So pleasing in itself, and to me so new," wrote Jones, it "could not fail to awaken a train of reflections" and "inexpressible pleasure," for it was

> fertile in the production of human genius, abounding in natural wonders, and infinitely diversified in the forms of religion and government, in the laws, manners, customs, and languages, as well as in the features and complexions, of men. I could not help remarking, how important and extensive a field was yet unexplored. (ix–x)

Field does fascinating double duty here, referring both to Jones's physical field of vision as his voyage neared its end—"*India* lay before us, and *Persia* on our left, whilst a breeze from *Arabia* blew nearly on our stern"—and to the fields of knowledge he imagined himself harvesting. In this dual signification, the word enacts Bacon's handshake—the meeting of the visible and intellectual worlds. Part of the changing "career" of world history, then, was to mediate worlds.

Since system, as we saw earlier, was well suited for that job, history's drive toward it played, and continues to play, a key role in readying historians for that same task. But this tectonic collision was not history's only career-altering encounter. In the early decades of the nineteenth century, history experienced at least three others:

1. Differentiation from the novel
2. The formation of the disciplines
3. The personification of agency

Each of these events sharpened history's professional profile, making it increasingly distinct from other genres and thus a more effective disciplinary tool. However, sharpening also entailed a narrowing that altered its relationship with system in a manner that bears importantly on choosing the best way for us to connect history to system now.

Differentiation from the novel. Like system, Jones's genre and his society's—the genre of "transactions"—proved to be a popular vehicle for history during the nineteenth century. But by the end of the second decade of that century, another genre—the one Fielding contrasted with system—became a particularly visible vehicle thanks to Walter Scott's lucrative turn from poetry to the novel. Note how this early review of Scott's first historical novel, *Waverley*, sets the scene of writing and reading in ways that clearly echo Jones's encounter with Asia:

> We are unwilling to consider this publication in the light of a common novel, whose fate it is to be devoured with rapidity for the day, and to be afterwards forgotten for ever; but as a vehicle of curious accurate information upon a subject which must at all times demand our attention—the history and manners of a very large and renowned portion of the inhabitants of these islands. … We would recommend this tale, as faithfully embodying the lives, the manners, and the opinions of this departed race. (*British Critic* 1814, 204)

We have become used to thinking of Scott as a matchmaker, as one of the first novelists to bring history *into* the novel. But history, as Cheryl Nixon's *Novel Definitions* illustrates in detail, was already there—there from the start when the genre of the novel first emerged from the categorical stew that featured romance as its other primary ingredient (2008, 44). Far from bringing the genres of history and the novel together, Scott's industry and popularity played a key role in writing them apart.

To use Samuel Coleridge's term, Scott's efforts and reception desynonymized history and the novel. Their relationship became adjectival, authorizing two different kinds of difference. First, dwelling on some novels as "historical" allowed for a category of those that were not. That left the novel—after more than century of being closely linked with history (a sharing enshrined on hundreds of title pages and in the quotation heading this chapter)—to stand on its own. Second, by the time the Romantic period, as we conventionally label the late eighteenth and early nineteenth centuries in Britain, drew to a close, reviews of Scott and his novels regularly, and often ferociously, focused on how history was different from what Scott wrote. In 1847, for example, *Fraser's Magazine* took a review of Scott as an occasion for asking whether "History gained by his writings":

> That a great and romantic effect was thus produced [in *Waverley*] is evident. There is all the semblance of a genuine historical *tableau*; the elementary characters are living, breathing men, and they offend us by no discrepancies of manner or costume. But is historical truth preserved? We confidently answer that it is not, and there is no surer way of contravening the realities of History.

This, then, was the consensus that emerged by midcentury: Scott, thanks to his strengths and weaknesses, became the poster boy for doing what the eighteenth century could not: *differentiate* history from forms of fiction such as the novel and the romance.

The formation of the disciplines. That differentiation, in turn, was part of the more comprehensive reclassification of kinds and knowledges that we have just tectonically identified as disciplinarity: the reorganization of knowledge into the narrow but deep fields we inhabit today. History was not alone in hiving itself off as its own form of specialized knowledge; every newly forming discipline instituted itself as a "detached" form of knowledge, to use *Britannica's* word, by telling its *own* history (Siskin 1998). In literary studies, for example, *Romanticism* is the label for the historical tales literary study tells itself about the period in which it became a discipline. During the late eighteenth and early nineteenth centuries, the first courses in English Literature were taught, the first departments of English were formed, the essay and the review—as well as the periodicals that contained them—assumed their modern forms, and our current disciplinary distinction between the humanistic and the scientific was first instituted.[9] History thus experienced a doubling of its status: its incarnation as a stand-alone discipline was coupled with its instrumental role in the formation of its freestanding neighbors.

The personification and hegemony of agency. The infrastructure to support history's specialization and professionalization was assembled during the last three decades of the eighteenth century. Once in the proximity of system, history—scaled up and with temporality at its core—was newly energized by the notion that history might itself have a story (Pomata and Siraisi 2005, 234). This disciplinary reflexivity bracketed the Romantic period, from the stadial histories of civil society—with their built-in stories of progressive stages—that surfaced in late eighteenth-century Scotland, to Gibbon and Tytler, to the very large-scale teleological and dialectical epics of Hegel and Marx in the second quarter of the nineteenth century.

As history's own stories of historical change stretched temporally and geographically from local reports on things as they are to universalizing tales of things as they were, are, and must become, gaps in the narratives were filled in two primary ways. "Ideas," which had been, in Donald Kelley's words, "rational and universally valid concepts independent of time," became, in the context of these new kinds of history, things that "lived in time"—and thus capable of explaining how times changed (Pomata and Siraisi 2005, 231). In retrospect, all that was necessary to give us that most familiar form of modern history—the "history of ideas"—was a companion for this new form of life: a subject capable of generating and carrying ideas forward in time. If, per the reading of Luhmann I cited earlier, system is now receiving the subject's mail, this was the earlier moment when letters first interpellated that subject. They hailed it into history as an agent personifying some of system's causal functions—functions that Luhmann's work has tried to reclaim for system.

That developmental subject—a subject now defined and made deep by the capacity to change *over* time—advertised its newly intimate relationship with ideas in a wide range of genres, from the philosophical—as in Kant's motto for Enlightenment, "dare to know"—to the lyric forms that scholars of the Romantic period know so well. But as William Warner and I have argued in *This Is Enlightenment* (2010), Kant's daring subject was not in need of Enlightenment; it was the *product* of Enlightenment, including, as we detail, Europe's first postal systems. By 1784, the year of Kant's famous essay on Enlightenment, man had already become, in Bacon's terms, a new kind of tool—a tool whose power now lay in its insistence on using its own understanding to change itself (Siskin and Warner 2010, 1–21).

Empowered in this autopoietic manner, this embodiment of agency amplified agency itself. It scaled up its own tales of personal development into a form of history *as* development, appropriating, simplifying, and thus amplifying the existing causal narratives—from the stadial to the gravitational to the universal. It thus played a key role in establishing the hegemony of agency over history, securing the assumption that history's primary task was to tell tales of causal relationships between past and present. The history of ideas became the dominant form of this causal history as its narratives were increasingly driven by two kinds of persons: personifications of ideas themselves, such as "capitalism," and individual persons whose own daring ideas changed history.

What the subject did to history, then, was give us both Marx and Carlyle. As ways of mediating history's relationship to system, they and their histories have become useful disciplinary tools, extending historical inquiry in innovative ways. However, those uses came with a price: history's encounters with differentiation, disciplinarity, and agency all generated the same vector. Even as they helped history to scale up in content, they narrowed it as a form, delimiting its interrelations with fiction and other fields as they sutured temporality and causality into a single prime directive. Narrowing, let me emphasize, is a relative term—my example in the next section is *On the Origin of Species*—and it is not in itself a problem, but it can become a problem when conditions of possibility change.

SYSTEM AND HISTORY II

To dramatize why this matters for finding the right histories for system now, at the moment of its new proliferation in new technologies, I will go out on a disciplinary limb. Actually, let's call it a branch, for my model here is the man who won a Nobel Prize for establishing that there are three basic branches to the tree of life. Instead of resting on his personal laurels and the recent prominence of biology among the sciences, however, Carl Woese has spent his seventh decade mounting a campaign in scholarly articles and magazine interviews to recast his field. Declaring the obsolescence of reductionist biology in the very moment of its apparent success, he has been arguing, in the physicist's Freeman Dyson's words, for "the need for a new synthetic biology based on emergent patterns of organisation rather than on genes and molecules" (Dyson 2006, 38).

Woese's enabling strategy is to return to the usual suspect in arguing about biology, Darwin, but not just to take sides in the debate that still concerns all of us, inside and outside the field: the debate between evolution and creationism. Instead, he zooms out, propelled by the simple but surprising observation that Darwinian evolution is only a chapter in the book of life. We forget that Darwin wrote about the origin of species, not the origin of life. Though the age of Darwinian speciation, roughly 2 to 3 billion years, may seem long to us, it was but an interlude within the larger history of living things. What preceded it was a period without species that Dyson has summarized in often hilarious terms:

> Life was then a community of cells of various kinds, sharing their genetic information. ... Evolution ... could be rapid, as new chemical devices could be evolved simultaneously [through] horizontal gene transfer. But then, one evil day, a cell resembling a primitive bacterium happened to find itself one jump ahead of its neighbours in efficiency. That cell separated itself from the community and refused to share. Its offspring became the first *species* of bacteria, reserving their intellectual property for their own private use. (Dyson 2006, 39, emphasis mine)

Once VGT (vertical gene transfer)—how species do it—replaced HGT (horizontal gene transfer), evolution slowed down, for "individual species once established evolve very little." In Darwinian evolution, established species have "to become extinct so that new species can replace them."

Strangely enough, that imperative sealed the fate of Darwin's epoch; it began to end

> when a single species, Homo sapiens, began to dominate and reorganise the biosphere. Now, as Homo sapiens domesticates the new biotechnology, we are reviving the ancient pre-Darwinian practice of horizontal gene transfer, moving genes easily from microbes to plants and animals, blurring the boundaries

> between species. … The evolution of life will once again be communal, as it was in the good old days before separate species and intellectual property were invented. (39)

So while almost everyone else thinks they are in on the big argument—the "truth" of Darwinian evolution—Woese has put that concept, and thus the controversy, in its historical place by proposing a different history and hierarchizing them: the history of life *includes* the history of species. And, not surprisingly, a crucial piece of technology for this scaling up is system. At the moment of Woese's call for a history of life, "systems problems," according to the Systems Biology website of Harvard University, "are emerging as central to all areas of biology and medicine." Although *systems biology* is a label for many different kinds of endeavors, it most often shares two key features with Woese's prescription for a biology newly adequate to a history of life: it is synthetic, requiring institutional and intellectual "synergisms" across the sciences, and it attends in particular to the modeling and discovery of emergent properties.[10]

Turning to Woese thus provides a current example of how the interrelations of system and history continue to change, particularly as our world (as in my title for this part of the book) becomes a world full of systems. There are as yet no new departments on the "systems biology" model outside of the hard sciences, though there has been a proposal for "systems sociology" at MIT, and much work, as we've seen, in systems theory and in particular kinds of systems across the humanities and social sciences. Woese's venture speaks to those disciplines as well, for although his immediate goal was to improve his own field, his method echoes and enacts Bacon's plan for advancing all knowledge through a careful choice of histories. What kind of history can do for the study of system what Woese's history of life offers to biology? Bacon's list of histories was long, but he did hierarchize it, and in doing so he singled out one history in particular as strategically capacious.

RECLAIMING LITERARY HISTORY

> History is Natural, Civil, Ecclesiastical and Literary; whereof the three first I allow as extant, the fourth I note as deficient. For no man hath propounded to himself the general state of learning … without which the history of the

> world seemeth to me, to be as the Statue of Polyphemus with his eye out; that part being wanting, which doth most shew the spirit and life of the person.
> —Francis Bacon, 1605 (2008, 175–176)

For anyone outside an English Department, the notion that "literary history" is the living eye of the "history of the world," would certainly raise eyebrows. Bacon identified one reason: this kind of history was, in his time, "deficient." That is not surprising given its remit as Bacon defined it: this "story of learning" would track "the antiquities and originals of knowledges … and all other events concerning learning, throughout the ages of the world" in order to "make learned men wise in the use and administration of learning." It was, in other words, "HISTORIA LITERARUM" in the formerly comprehensive sense of letters as all written records. Through the late eighteenth century, *letters*, *learning*, and *literature* remained interchangeably inclusive.

"Literary history" continued to be deficient—there in "divers" parts, as Bacon put it, but not in a whole from "age to age"—until the meaning of the word *literary* changed. As "Literature" became a subset of its formerly inclusive self, literary history became less about knowledge in general and more about the narrowing of knowledge into disciplines. It then followed exactly the vector I described for "history" as a genre at that time. Its interrelations with system precipitated various classificatory schemes, from genres to periods, but what came to occupy the center was the subject. That self made itself at home as the primary causal agent of literary history: the Authors who still populate anthologies and remain principal objects for articles and books.

After two centuries of equilibrium—a balance between the scale of its content and its authorial patterns of organization—literary history in the late twentieth century began again to be experienced as deficient. Its field began to change shape—both deepening, into searchable but growing databases of historical and current materials, and extending, across the globe and into and through new technologies. What had been intimidating to specialists in training—from Beowulf to Virginia Woolf—had quickly become a sliver of what was easily accessible.

Resistance to scaling up soon gave way to the need to choose how. One alternative has attempted to conserve the discipline—and its relationship to its authorial subjects—by growing the existing schemes (e.g., expanding the canon, consolidating and adding periods).[11] The obvious problem has been determining how much is too much for schemes designed to handle less. And especially troubling in that regard has been the effectiveness of the tools that do the handling. The tools of literary history have primarily been tools of scarcity; they were designed, that is, to make more out of their objects.

Consider one of the most popular tools: close reading. From the moment at the end of the eighteenth century when the poet William Wordsworth told the readers of *Lyrical Ballads* to read his poems in the same "style" that he wrote them—"look[ing] steadily"—writers and readers of literary history have moved in closer to the text in order to make more out of it (Wordsworth [1800] 1974, 1:132). The further they have burrowed in—isolating each word, each image—the more they have had to say. Canonization, the vector of less, and close reading, the vector of more, have been the paradoxical countermovements of the modern literary system. But that system's tools, like all analytic tools, are data sensitive; they work differently at different scales.[12] Thus, as the canon has grown, canonization has ceased to be the vector of less, calling into question the value of close reading's extractions of more.

This is how a narrowing that was once enabling has become a problem. If literary history can't scale up while conserving the literary system it helped to form, then it's Woese time: time to find a new history that can contain the old one. For biology, this meant a history with a different name—"life" as embracing "species." But Woese time also points back in time—to Bacon as the strategist who first linked advancement to a hierarchy of histories and put literary history in the top tier. To reclaim his earlier, broader sense of *literary* is to acknowledge our own historicity—the aging of our deep but narrow disciplines and the issues of scale they now face. But reforming literary studies is not my agenda here. Though I would welcome that as an effect, my goal in connecting past to present is to focus on system.

To that end, this book as I ventured in the Prologue is best described as a literary history in the Baconian sense that I am reclaiming here. As a story of "knowledges," it is compatible with both the book's ambition—of describing how system has shaped knowledge—and its strategy of including histories of other kinds. That strategy can best be understood as applying the lessons of *historia*'s past—that history has consisted historically of different kinds with different functions—in a contemporary way. The histories I am about to describe, and then use, "run" in this book like programs multitasking on a computer. Each one can perform particular tasks, some tasks enable others, and they all relate hierarchically to each other. While systems within systems, as we have encountered them so far, have highlighted that genre's architecture as scalable, histories within histories, as they will be deployed here, illustrate the ways they can be interoperable. Some histories, for example, can be ways of doing other histories.

THE HISTORY OF MEDIATION

Bacon did not leave a model for how to do his version of literary history, and it's not my purpose to offer a prescriptive one. The value of flagging this book's genre as literary history derives from the act of grouping itself. Sorting out Wallerstein and Luhmann, historicizing ideas and subjects, and connecting Woese to Bacon are all classificatory acts that yield a distinctive mix of features. Those features highlight scope, interrelations, and change, and they are formal ("literary" as all written records, including multiple histories), thematic (knowledge of all kinds), and temporal (changes in knowledge over time). Together, they constitute a genre adequate to the ways I engage system: a literary history of system studies system as both a genre that interrelates with other genres and as a technology that shapes and reshapes knowledge.

But how? What kinds of history can do the work of this form of literary history? Conventional literary history has largely depended on its interoperability with the history of ideas: authors have ideas that they represent and then critics interpret them. What would an alternative currency be, and where can we find it? The opening words of Bacon's *Great Renewal* provide some explicit directions: "The human intellect is the source of its own problems, and makes no sensible and appropriate use of the very real aids which are within man's power" (Bacon[1620] 2000, 2).

Outside the mind and its ideas are what Bacon called "aids"—one of a cluster of terms and translations that includes "machines," "instruments," and "tools." Whether moving a "heavy obelisk" or "advancing" knowledge, relying on the mind's own ideas was, Bacon concluded, an "act of utter lunacy" (28–29). Sanity, for Bacon, was accepting the necessity of tools—tools that work. His "renewal" of knowledge was thus fundamentally a problem of what William Warner and I (2010) have termed *mediation*. We use that word as shorthand for the work done by tools, by what we would now call *media* of every kind—everything that intervenes, enables, supplements, or is simply in between—emphasizing the Baconian stipulation that media of some kind are always at work.

If they are, then that work can be a framework for identifying and articulating change: mediation never ceases, but the forms of mediation differ over time. A history of this kind could be a useful alternative to the history of ideas. In *This Is Enlightenment*, Warner and I posit mediation as the inclusive term for this history; mediation can refer to what we now call *media*, but (as the breadth of reference of that book demonstrates) it is not restricted to them. The history of mediation can thus engage "media history" and "media theory," but its wide range of objects, forms, technologies, agency, and interactions—and thus its chronological scope—differentiates it from both of those established enterprises.

Our test case for this history was to present Enlightenment as "an event in the history of mediation," and we found that it provided new, and newly useful, ways of contesting the distortions of the history of ideas. Once ideas are animated with the requisite agency to change history on their own or through designated subjects, good historiographic intentions fall victim to the Frankenstein syndrome: once you start them, you can't stop them. Thus, the Frankfurt School chased their marker for Enlightenment—"instrumental reason"—from Homer to Hollywood. And for those uncomfortable with such chronological excess, the history of ideas has also conveniently stretched in the other direction. Instead of using one idea to extend a single Enlightenment across thousands of years, for example, it has turned different ideas, within more conventional time frames, into multiple Enlightenments, from the Counter to the Radical to the Postcolonial.

Viewed as an event in the history of mediation, however, Enlightenment doesn't dissolve into too many years, too many things, or too many versions of itself. Because mediations can be more easily pinned down than "ideas" to specific times and places, we can track more of them more accurately—including genres such as periodicals, infrastructure such as schools, associational practices such as clubs, and protocols such as copyright—and thus more readily identify patterns in those interactions. And since mediation doesn't force us to distinguish between living things and inanimate things—it does not discriminate, that is, against any particular form of agency—it points us past the increasingly unproductive binary of technodeterminism. A history of mediation thus provides new, and newly useful, ways of thinking about change. And by reversing the narrowing of Romantic historiography, it gives us the capacity to zoom out Woese style—and thus write a literary history of the scope and scalability that system needs.

Such a history must also be able to take into account system's capacity to adapt to different conditions and substrates, from writing and print to the algorithmic and the electronic. As an alternative to the focus on representation and interpretation in the history of ideas, mediation can ask the question "*in* what?"—and thus "capture," as John Guillory has argued, the "hidden complexity of the process" that representation has for so long purported to describe, particularly the issue of "'in what' form a representation is transmitted." Mediation thus offers an alternative to what Guillory calls "the dominance of representation in Western thought" (Siskin and Warner 2010, 7).

Using a history that addresses mediums is crucial to engaging system as a genre, for genres are themselves forms of mediation. Ideas never float free of form but are always *in* genre; they are never unmediated. Classifying into genres, as I have been doing with history in this chapter, is also an act of mediation—both between things (by generating and identifying hierarchical relationships of sameness and difference) and between us and things. In the *Fraser's Magazine* discussion of Scott, the novel is described as a "medium" for history.

With its focus on substrates and conditions, of the physical embeddedness of ideas—they are always already physical—mediation also provides a way past some of the pitfalls not only of the binary of technodeterminism but also of the binary that the history of ideas always invites: idealist versus materialist and the causal arguments it generates. I am not suggesting here that mediation resolves all of those arguments given the longstanding and important investments that we have in them. But I do hope this book gives a sense of what might be gained by turning to mediation. I see those gains as taking the form both of specific insights—as with the arguments about Enlightenment sketched out above—and of a more general and much-needed push toward compatibility between historical work and a growing consensus in other disciplines on the relationship of information and knowledge to physical reality.[13]

Having made these claims for mediation, I should also reiterate here that choosing the history of mediation as a way to operate literary history at Bacon's scale of stretching from "age to age" cannot be the task of one book. Since this one turns back to Galileo and forward to the fate of system today, I recognize that I am risking criticism for what will inevitably be left out in stretching as far as I do. *Far* is, of course, a relative term. The book that shares most clearly with mine the task of Baconian literary history stretches even further. In *A Social History of Knowledge* (two volumes published separately in 2000 and 2012), Peter Burke scales up both chronologically and geographically—a strategy he explains in terms of both disciplines and genres. "Relatively few historians," he points out, "have taken the sociology of knowledge seriously." His notion of seriousness echoes, as I just have, the choice that Samuel Johnson posed for eighteenth-century writers. This particular history, Burke proclaims, is

> an essay, or series of essays, on a subject so large that any survey which did not take a consciously provisional form would be not only immodest to attempt but impossible to carry out. I must confess to a predilection for short studies of large subject, which attempt to make connections between different places, topics, periods or individuals, to assemble small fragments into a big picture. (Burke 2000, 9–10)

In terms of Johnson's opposition, Burke's essay is his "consciously provisional" alternative to a complete system.

It is in the trade-offs of scale—"short" and "large," "small" and "big"—that his effort complements mine. I use *complement* in the strongest sense; for the reader this means engaging our endeavors not as rivals to be judged against each other but as contributions to the joint enterprise of a newly capacious literary history. We share, to offer one example, the same primary answer to the question, "Whose knowledge is the subject of this study?" The emphasis, in Burke's words, "following the sources, [falls] on dominant or even 'academic' forms of knowledge, on 'learning' as it was often called" in the past. But Burke also essays a "wider" framework to which I have not aspired in this book: "The competition, conflict and exchange between the intellectual systems of academic elites and what might be called 'alternative knowledges' will be a recurrent theme in this study" (14). The price of accommodating that theme, however, is the narrowing of other parts of the story of knowledge—such as the treatment of "system" itself. Burke provides a paragraph on the "coming into use" of system in the seventeenth century in the first volume (Burke 2000, 86–87) and two paragraphs on its appearance as a "keyword" in the eighteenth century in the second (Burke 2012, 255–256).

I hope our readers will in the end conclude that the gaps opened by our particular negotiations of the essay/system binary are justified by the connections our studies do make. But those negotiations should be as transparent as possible, so let me clarify a few of the key temporal and geographical compromises—in the most productive sense of the term—that I have knowingly made. System's role in the Enlightenment is a central focus of this book. After the early seventeenth-century sightings we have just made, I turn to the long eighteenth century and its equally long linear rise of references to system. It is then, particularly in the century's middle and late decades, that the mediation of knowledge in general reached an unprecedented pitch. System, I will argue, played a key role in this extraordinary moment—a moment in which the very medium of mediation—its architecture of forms and tools, people and practices—became load bearing. At that point, each individual act of mediation worked not only on its own terms but also as part of a cumulative, collaborative, ongoing enterprise.[14]

My argument, as I sketched it out in the Prologue, then makes a number of stops in the nineteenth century to engage such topics as the emergence of the disciplines, Romanticism, and liberalism. It then returns to Darwin, not—as with Woese—to scale his work down, but to identify the major role his algorithmic use of system is now playing in current (nonmetaphorical) efforts to explain the universe as a computational system. I regret the many moments, topics, and uses of system missing from this agenda, but I take some comfort in the company of the collected volume whose title places it so close to mine in the enterprise of Baconian literary history. In his introduction to *The Shapes of Knowledge from the Renaissance to the Enlightenment*, coeditor Donald Kelley acknowledges that the "patterns and transformations" that should be part of his book

> cannot be well understood from the heights of the general history of ideas. … The map must be filled in by particular explorations and soundings, and our project called for a conference that would combine some encyclopedic (as well as interdisciplinary and international) breadth with scholarly and technical depth. (Kelley and Popkin, 1991, 1)

My book shares not only this concern about the history of ideas but also the sense of risk in attempting to construct an alternative.

Though writing on my own,[15] I too have tried to reach across and into other disciplines, and, at select moments, I have crossed national boundaries However, the focus of this book is Britain. Being more geographically capacious does, of course, have advantages, and an increased catchment area can alter some conclusions. For that I apologize, especially where my ignorance is the major factor. But sacrificing the advantages of following through in Britain would be cause for equal regret. It is too easy to lose arguments that do matter to the weight of limitless detail and qualifications, and to untempered admiration of the "trans-" and the "comparative."

The history of mediation is a particularly useful tool for navigating these hazards. In *This Is Enlightenment*, it spared us

> intellect-wasting custody battles over Enlightenment: "It's French, of course." "No, it's actually British." Grounded in specific mediations and yet yielding regularities, the history of mediation can clarify both the singularity of each local event *and* what those events have in common. (Siskin and Warner 2010, 11)

My more local focus on system has served up a number of compelling singularities: nation-specific arguments that are valuable in themselves but have the potential as well to be generative points of comparison:

- *System and the new science in England.* Bacon and Newton helped to enable Enlightenment by engaging (and not engaging) system in specific and influential ways. Bacon's plans for a new kind of knowledge used system to highlight difficulties in advancing knowledge. Newtonians were indebted not only to Newton's ideas but to the form of guesswork he choose to employ—a form that came to stand for the entire work. The *Principia* became known as the "Newtonian System."

- *System and the project of Enlightenment in Scotland.* By *project*, I mean that for Adam Smith and his Scottish cohort, Enlightenment was a carefully planned foray into the newly expanding world of print. Their systems, as we will see in *The Theory of Moral Sentiments* (Smith [1759] 1976), shaped knowledge in a specifically cumulative way that I describe as "Master Systems."
- *System and print in Britain.* The fate of Enlightenment in Britain was tied to the fate of system. To study system as a genre is to connect it and knowledge to the specific quantitative contours of the output of print in Britain.
- *System and politics in the United Kingdom.* The politics of a kingdom in the process of uniting offers, as we shall see, an exceptional opportunity for studying the history of system as something to blame, starting with the fact that the social and political body that emerged from the Glorious Revolution (1689) became known as "the English system."

Focusing through this national lens makes these particular uses of system visible. Sighting them in this way then sets the stage for future efforts to discover how system may have worked in similar and dissimilar ways elsewhere. To that end, I have engaged in some strategic boundary crossings, starting with the opening pairing of what Galileo saw and said with the simultaneous surfacing of the word *system* in English. I also integrate a number of the touchstones of system in France into my histories and arguments, from placing the distinction between *esprit de système* and *esprit systèmatique* into a history of blame to bringing Pierre Bayle's *Dictionnaire Historique et Critique* and Diderot's *Encyclopédie* into my narrative of system's role in forming the modern disciplines. Additional turns to such matters as Condillac's emphasis on system's ratio of principles to things and Rousseau's plotting his *Confessions* as a journey from system to system will also, I hope, help to map out avenues that comparative studies might take.

"PARTICULAR HISTORIES"—THE HISTORY OF BLAME AND THE HISTORY OF THE REAL

> If I ask him to out to dinner, a race, rout, or ball, he replies, "You may go, if you please, but it is not in my system."
> —*The Systematic, or Imaginary, Philosopher: A Comedy in Five Acts* (1800)

Histories can work like camera lenses, capturing an object from different angles. The key to taking the most telling snapshots lies in selecting the histories into which the object is written. At the end of the *Novum Organum*, after describing his "Mother History"—the "Natural and Experimental History" that he envisioned as the primary vehicle for advancement—Bacon added the long list of what he called "*Particular Histories*." He saw those histories as legal briefs that would prepare the "human race" to "cross-examine nature herself" (Bacon [1620] 2000, 223, 232). For my examination of system, two histories in particular work in and across the arguments to follow: the history of blame and the history of the real. The quotation at the start of this section is a punch line for both.

The philosopher is the butt of this comedy because of system. Two hundred years after Bacon worried that system could distract people from things as they are, this philosopher is distracted from almost everything. His experience of the world is thoroughly mediated by system; in fact, he is *in* system. To mock his behavior is to put system itself in the history of blame—and to write that history is to prepare a brief on one of the major ways that system has worked in the world.

Early in my research for this book, I discovered that it continues to work that way. In an effort to enlist my students in the task of identifying when and how system first became an object that could be blamed, I constructed an entire undergraduate course called "Blaming the System." A day after the English Department submitted its schedule to the university, I received an e-mail from the registrar asking if she could take the class. Any sympathy that note evoked in me, however, completely evaporated a few weeks later when the next semester's bulletin emerged from her office: my course title had been mysteriously transmuted into "Blaming the Victim." I still can't imagine what the students who signed up for the class were thinking when they first entered the room, but within a few weeks, they had done some rewriting of their own. EGL 347 became "Blaming the Siskin."

What the class helped to uncover was a startlingly wide range of contexts in which system has been blamed. As we shall see in the chapters that follow, these range from politics to economics, from religion to epistemology. Blaming was so pervasive, in fact, that it doubled in on itself, with some people blaming others for habitually blaming the system: "If the Great Mind had form'd a diff'rent Frame," wrote Sir Richard Blackmore in 1712, "Might not your wanton Wit the System blame?"

Although often mocked, as it is here, for excess, blaming the system played a crucial role in shaping knowledge. Blame so often accompanied system because systems were seen as something made—made to approximate the real and thus always falling short of it. But opening a gap invited more knowledge to fill it, and thus system configured the space in which Enlightenment *could* take place. Blaming the system set the scene of Enlightenment.

That scene was also a scene in a different "particular history"—the history I call the "history of the real." The "systematic" philosopher was also, per the play's title, the "imaginary" philosopher in at least two ways. Being "in" system detaches him from the world that is real to those who mock him—the world of diners, races, routs, and balls. But he is also an imaginary philosopher in the sense that he is not really a real philosopher but a fake one trapped by system. System is in some way an index to both what is experienced as the real and what can be ontologically claimed as real.

We have already seen this connection in Galileo's account of Copernicus as "taking off the clothes of a pure astronomer" in order "to investigate what the system of the world could really be in nature." The history of the real provides a frame for considering continuities and discontinuities in what the real could really be. Getting its bearings from objects such as systems, this history is an effort to identify how and when the experience of the real—not just opinions about the real—changed. For the eighteenth century, for example, consider all of the discussion of *tabulae rasae*, primary ideas, probability, fiction, fact, novel versus romance, imitation, falling trees, kicking rocks, skepticism. These phenomena all point to the same historical fact: the real was up for grabs during this time period, and what people were grabbing at was the physical.

When did the physical become real? When, to put this in more philosophical terms, did the physical secure its ontological hold on the real? For Newton, the world was still metaphysical. Soon after, however, the physical started kicking, becoming the virtual reality of the late eighteenth century; that is, the physical at that point was very present but still experienced as an approximation—something that was not quite real. Briefed by the history of the real, we can query system about these changes. At different times, system has been the metaphysical form of the divine will, the physical form of knowledge in print, the virtual form of our electronic simulations, and, now, the informational form of the universe itself. As we move in Part II forward into the Enlighten-ment—and, in the Coda, into the future—we will see how system has shaped and reshaped our knowledge of the real—and thus our experience of it.

Part II MEDIATING KNOWLEDGE

SYSTEM AND THE FATE OF ENLIGHTENMENT

3 THE PROJECT OF ENLIGHTENMENT (MASTER SYSTEMS)

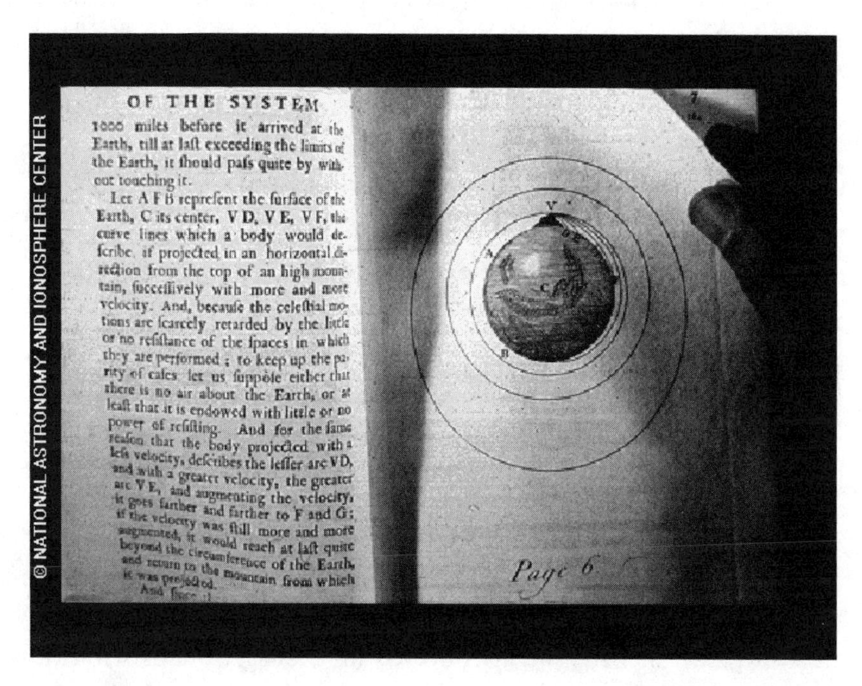
OF THE SYSTEM

1000 miles before it arrived at the Earth, till at last exceeding the limits of the Earth, it should pass quite by without touching it.

Let A F B represent the surface of the Earth, C its center, V D, V E, V F, the curve lines which a body would describe if projected in an horizontal direction from the top of an high mountain, successively with more and more velocity. And, because the celestial motions are scarcely retarded by the little or no resistance of the spaces in which they are performed; to keep up the parity of cases let us suppose either that there is no air about the Earth, or at least that it is endowed with little or no power of resisting. And for the same reason that the body projected with a less velocity, describes the lesser arc VD, and with a greater velocity, the greater arc VE, and augmenting the velocity, it goes farther and farther to F and G; if the velocity was still more and more augmented, it would reach at last quite beyond the circumference of the Earth, and return to the mountain from which it was projected.

And since [illegible]

Figure 3.1
Image from *A Treatise of the System of the World* onboard *Voyager* 1.

What's at stake in tracking system can perhaps best be captured by the image in figure 3.1. It scales up that genre's history by adding new temporal and spatial dimensions to its trajectory. As I was writing about it, this image has become one of the first human artifacts to leave the solar system. To put this more succinctly, the system of the world is about to leave our system. In 1977, NASA chose pages from Newton's *A Treatise of the System of the World* to represent humanity to whomever or whatever else is out there, encoding this image of those pages onto a twelve-inch gold phonograph record packed into the *Voyager* 1 spacecraft. But there's a twist to this apparently simple act of homage to the *Principia* and its author. These are not pages from the *System of the World* famously published as Book 3 of the *Principia*. It is a very, very different version—one that Newton rejected; it was, in fact, not even published until 1728, the year after Newton died.

And there's another twist to that twist. Newton came very close—repeatedly—to sending his treatise into the world without any "System" at all. Why was system such a vexed issue in the late seventeenth century? And why, once it did make it onto the printed page, did *system* become so successful that it was launched into space as one of humanity's calling cards three centuries later? How did that particular form of knowledge come to represent our species' ability to, as Francis Bacon put it, "advance knowledge"?

For Bacon, the key to advancement, what he called

> the whole secret is never to let the mind's eyes stray from things themselves, and to take in images exactly as they are. May God never allow us to publish a dream of our imagination as a model of the world. (Bacon [1620] 2000, 24)

Even editors who do not experience their work as a religious vocation may find themselves occasionally haunted by this last sentence—"May God never allow us to publish a dream of our imagination as a model of the world"—just as the first sentence, with its imperative to see "things as they are" has haunted historians of science, eliciting tales of empiricism and induction.

To test those tales, let's try an experiment. Here are some "things" to look at. Figure 3.2 is a recent photograph of what Galileo saw through his new spyglass—the things that conveyed his message. But as with the image on *Voyager*, there's a twist to this image as well: the message from the stars was that these things were not stars at all—they were moons revolving around Jupiter. This next image, figure 3.3,[1] is what we would call stars—Galileo would have called them "fixed stars."

Look closely. Can *you* see these things *as they are*?

For the physicist David Deutsch, this is an epistemological question with a historically specific answer:

> How do we know? … No human has ever been at the surface of a star. … Yet we see those cold dots in our sky and know that we are looking at the

Figure 3.2
Jupiter with four of its moons.

Figure 3.3
"An earth's-eye view straight through the disk of the Milky Way" from the Calvin Observatory. Courtesy Brad DeFrain, Calvin Observatory.

> white-hot surfaces of distant nuclear furnaces. Such ideas do not create themselves, nor can they be mechanically derived from anything: they have to be guessed. (Deutsch 2011, 3)

Four hundred and one years after Galileo broadcast his message from the stars, Deutsch published an even briefer one: they are *nothing* like they look. The difference between "dots" and "white-hot surfaces of distant nuclear furnaces" is the product of what Bacon called for—the advancement of knowledge. It is also the informing irony of Bacon's plan for the great renewal: the reward for seeing "things" as he demanded—"exactly as they are"—was to learn to see them in an entirely different way. In the history of mediation—the history, as I have been deploying it, of the forms and tools that mediate knowledge—that irony took the form of a substitution. One form of mediation had to be replaced by another. The imaginary models that Bacon thought had stalled knowledge—what he also called the "idols of the mind"—had to be swept aside so that things could be worked on with a new set of instruments. That work has been guesswork.

The historical importance of guesswork since the seventeenth century helps to account for what Deutsch (2011) calls "one of the most remarkable things about science": "the contrast between the enormous reach and power of our best theories and the precarious, local means by which we create them." Guessing, of course, is an act that can occur at any time, but the work it performs depends on both context and form. Bacon's emphasis on things was not a simple turn to empiricism for the raw materials of induction; it was part of a strategy for changing contexts that cleared a space for the new tools of guesswork. What was cleared away was the scholastic reliance on both authoritative texts and the forms by which that authority was exercised.

The two primary ways of exercising that reliance were commentary and questions. The former extrapolated from the text through forms of logic and method. The latter used those same tools to take the scholar or student back from the question to the text because the answers were there. Neither things nor guesses interrupted these circuits, which made them efficient forms of pedagogy for these schoolmen. Two genres in particular were important vehicles for circulating knowledge through those circuits: tractates and summas.

During the late Middle Ages, as Edward Grant (2007, 155, 288) has documented, a number of scholastic authors had written tractates (*tractatus*), or treatises on specific themes. Through the sixteenth century, these were often written as introductions to issues primarily drawn from Aristotle but sometimes on topics that had been introduced into natural philosophy by other commentators. A treatise that summarized a particular topic was a summa, but summas could also take the form of a set of questions. Answers to those questions collectively covered the various parts of a topic by securing them back to the text from which the questions were generated.

By the seventeenth century, increasing numbers of treatises began to cut the interpretative and interrogative ties back to scholastic forms of authority. The systems that first emerged among Ramists such as Johann Heinrich Alsted and Bartholomew Keckermann played a transitional role here. As a form used to feature relationships between parts and wholes, system at that time lent itself to these logical extrapolations of scholasticism. Some of the ambiguity we find among later writers toward the genre of system has its origin in this early link to the scholastic. Adam Smith, for example, used system extensively, and, as we shall see, he even declared his own and everyone else's "love" for it. But he also projected this particular use of system back onto a Stoic philosopher he wished to blame for reducing

> their doctrines into a scholastic or technical *system* of artificial definitions, divisions, and subdivisions; one of the most effectual expedients, perhaps, for extinguishing whatever degree of good sense there may be in any moral or metaphysical doctrine. (Smith [1759], 1976, 185, 291, emphasis mine)

But system, like treatise and other genres of knowledge, could be used to perform other kinds of work. A genre can vary its features and functions because genres are groupings that change over time—they are empirical, not logical. Once the turn against Aristotle and toward "things" deflated the value of scholastic diagramming and sets of questions, the feature of parts and wholes was repurposed to what I am calling guesswork—to guessing, for example, which parts belonged to which wholes.

By 1728, in his *Cyclopaedia*, Ephraim Chambers highlighted precisely that feature in defining system as:

> the Whole of any Doctrine, the several Parts whereof are bound together, and follow or depend on each other. (Chambers 1728, 2:165)

Definition, however, was only the start for Chambers. His "universal dictionary" also promised "accounts of the things signify'd thereby." Thus a few lines deeper into the system entry, Chambers launches the reader on a quintessentially cyclopedic journey through the early eighteenth-century landscape of new knowledge. The first stop in this cross-referencing—a feature Diderot enthusiastically adopted for his encyclopedia—links system to another form of guessing:

> System and Hypothesis, have the same Signification; unless, perhaps, Hypothesis be a more particular System; and System a more general Hypothesis. See Hypothesis.

If we circle back as ordered to volume 1 of the *Cyclopaedia*, we find *hypothesis* defined as

> a kind of System, laid down from our own Imagination, whereby to account for some Phenomenon or Appearance of Nature. (1:281)

By inserting "Imagination" into his definition, Chambers echoed Bacon's fear from a century earlier—that we might "publish a dream of our imagination as a model of the world." That worry then becomes the occasion for neatly summarizing the debate over the shape of knowledge at the moment that the event we now call Enlightenment began:[2]

> The real and scientific Causes of natural Things generally lie very deep: Observation and Experiment, the proper Means of arriving at them, are in most Cases extremely slow; and the human Mind impatient: Hence we are frequently driven to feign or invent something which may seem like the Cause, and which is calculated to answer the several Phaenomena; so that it may possibly be the true Cause."

"Such Fictions or *Hypotheses*," Chambers concluded, "are much less current now than formerly," for philosophers had become "divided" as to their use. He then clarified his own position by asserting that "the latest and best Writers are for excluding *Hypothesis*, and standing wholly on Observation and Experiment."

Not surprisingly, given the frequency we have already noted of the "Sir" Isaac" "Newton" wordplates on title pages at that time, Chambers then invokes the author of the *Principia*:

> Whatever is not deduc'd from Phaenomena, says *Sir Is. Newton*, is an *Hypothesis*; and *Hypotheses*, whether Metaphysical, or Physical, or Mechanical, or of occult Qualities, have no Place in Experimental Philosophy.

This statement is from the penultimate paragraph of the "General Scholium" added to the end of 1713 second edition of the *Principia*. It was thus an echo of how Newton had chosen to open the *Opticks* nine years earlier (1704): "My design in this Book is not to explain the Properties of Light by Hypotheses, but to propose and prove them by Reason and Experiments" (Newton 1952, 1). Hypothesis was in such ill repute by that second decade of the eighteenth century that Roger Cotes opened his editor's Preface to the *Principia*'s second edition with a full-scale attack on those who "take the liberty of imagining … whatever they please":

> When they do this, they are drift off into dreams, ignoring the true constitution of things, which is obviously to be sought in vain from false conjectures, when it can scarcely be found out even by the most certain observations. Those who take the foundation of their speculations from hypotheses, even if they then proceed most rigorously according to mechanical laws, are merely putting together a romance, elegant perhaps and charming, but nevertheless a romance. (Newton [1713] 1999, 385–386)

If *system* had been functionally synonymous with *hypothesis*, then it too would have been banished into romance and from a central role in Enlightenment. But the consistent and telling qualifications in Chambers's entries presage a different tale. Read together, they describe how a conceptual and functional wedge formed between the two genres that angled system into its linear rise.

In his entry for *system*, Chambers inserts the wedge by adding a second thought to the initial claim of hypothesis having the "same Signification." The turn is tentative—"unless, perhaps"—but it does pry the terms apart with the suggestion that they operate on different levels of particularity. That gap widens in the *hypothesis* entry with the assertion of an essential difference in their similarity; system here is hierarchically dominant with hypothesis but one of its "kind." The cyclopedic turn back to "system" then drives the wedge even deeper—deep enough to open a historical split between them:

> Some late Authors, indeed, furnish a fresher Distinction: An Hypothesis, say they, is a mere supposition, or Fiction; founded rather on Imagination, than Reason: a *System* is only built on the firmest Ground, and raised by the severest Rules.

The furnishing of this "distinction" proved to be a crucial adjustment to the historical trajectory of system, functioning like the controlled burns that change the attitude of a spacecraft, redirecting and prolonging its flight.

Systems appeared in the skies and on the page at the start of the seventeenth century as part of the collective cutting of scholasticism's ties to authority. As a form for positing how different parts might combine into different wholes, system joined other genres in facilitating the necessary task of guesswork. I emphasize "necessary" to acknowledge Deutsch's correction to the standard histories of science as empiricism's enabling of induction (Deutsch 2011, 1–10). The path from empirical "things" to explanations—from seeing dots to knowing about furnaces—required guesswork, and that is precisely why Newton, Cotes, and Chambers all worried about how that guesswork should be performed. And we can read that worry in the historical use of—and comments on—the genres that performed that work. That's why system can tell us a great deal about the shape of knowledge from the Enlightenment.

One message that system conveys over and over again is that knowledge during the Enlightenment became a matter of doing things over and over again. Guesswork needed "things as they are" not because things like dots somehow suggest nuclear furnaces, but because things can help us to *choose between* guesses. After choosing, we can then guess and choose again, and then again, with the expectation that every iteration mediates the previous ones to produce a tighter fit between things and explanations. For Deutsch (2011), this process of error correction informed the "tradition of criticism" that distinguished Enlightenment, enabling its "sustained, rapid growth of knowledge" (13).

Here, then, are two important reasons why system played a central role in that tradition. Not only did its parts and wholes lend themselves to guesswork—to choosing what parts fit together into what wholes; their scalability, as we saw in chapter 1, proved to be scaffolding for error correction, as explanations were made to fit through adjustments to scale. Thanks to these specific features with these specific functions, the genre of system was increasingly experienced during the eighteenth century as sharing guesswork with other genres but potentially doing that work better. In Chambers's terms, knowledge shaped by system could be "built" and "raised" on the "firmest ground." The conditional is crucial here. Many systems could just as easily—to use a favorite word for failed systems we encountered earlier—"explode." There was no set way to build them and no guarantee of the results, but what mattered was not their fallibility but that they remained on the strong side of Chambers's "fresher distinction."

Ruminating in his commonplace book seventy years after the *Cyclopaedia*, Francis Jeffrey, one of the founders of the *Edinburgh Review*, still echoed Chambers's twists and turns at the very end of the century. "I really cannot help suspecting," Jeffrey begins, that "theory," "system," and "hypothesis" all share a "real meaning," but he then shifts immediately to delineating differences in how we use those words. While with *theory* we "attend" to "general views or principles," and with *hypothesis* we attend to "assumption[s]," with *system* we "consider more particularly the collection of facts that are presented to us." Here, again, is Chambers's "firmest ground" in contrast to the ungrounded guessing of system's partners. Confirming the familiarity of system by 1800, but also its plasticity, Jeffrey concludes with the casually phrased observation that "there is little difficulty about *systems*, tho' it is a pity the word should not have some more determinate signification" (Pitre 1980, 86–87).

If system was solely an idea—and thus identified and valued by what it *meant*—then the lack of a "determinate signification" might be a basis for regret—or for me, attempting to offer a more determinate definition in this book. But as a *genre*, system's flexibility of features and functions was a reason that systems had become so familiar and could be deployed with "little difficulty." It is what allowed Adam Smith, for example, to complain about systems while writing more of them. People wrote systems to avoid the very problems for which system was often blamed. And that meant more and more systems.

To sight those systems within the history of mediation is to see that they proliferated in part because that genre's flexibility and scalability allowed it to mediate itself—and thus efficiently enact the iterations of guesswork. In the history of ideas, however, the hegemony of agency figures that twist in a very different way—as a love/hate relationship among personifications. In Ernst Cassirer's romantic version, desire for freedom redeems system from blame:

> In renouncing, and even in directly opposing, the "spirit of systems" (*esprit de système*), the philosophy of the Enlightenment by no means gives up the "systematic spirit" (*esprit systèmatique*); it aims rather to further this spirit in another and more effective manner. … Enlightenment wants philosophy to move freely and in this immanent activity to discover the fundamental form of reality, the form of all natural and spiritual being. (1955, vii)

Whatever designs "Enlightenment" may have had on "philosophy," what those who wrote and used systems needed was clear and concrete: the proper genres for their undertaking. In system, they found and made a firmer form for the guesswork that could distance them from the determinacies of scholasticism.

THE NEWTONIAN DIFFERENCE—SYSTEM AND THE REAL

In his plan to jump-start that effort, Bacon feared that forms like system might easily be deployed using scholastic methods, with rhetoric[3]—rather than "things"—filling out the initially "empty" spaces. His solution was generic: claiming to follow the "earliest and most ancient investigators of truth," he wrote aphorisms. These "short, unconnected sentences, not methodically arranged," conveyed in their very form, he argued, the sense of knowledge in growth (Bacon [1620] 2000, 71). Another formal solution was the "essay" understood as an attempt or a fragment—what Samuel Johnson called a "loose sally." The plan was to start with these short forms and then proceed into the long list of histories addressed in the previous chapter.

On these choices of genre, insisted Bacon, hung no less than the fate of mankind. This may sound hyperbolic, but keep *Voyager* in mind as you listen to his prophecy. A process would begin, claimed Bacon, with

> such an end perhaps as in the present condition of things and the present state of thought men cannot easily grasp or guess. It is not merely success in speculation which is in question, but the human situation, human fortune and the whole potential of works. (Bacon [1620] 2000, 24)

For system to play a role in this new, high-stakes world of knowledge—not only to contribute to success in speculation but to become a track on

Voyager's phonograph record—it had to overcome Bacon's skepticism to gain Chambers's "firmest ground." This entailed new connections, not only the iterative ones to "things," but also social links to new configurations of knowledge. Guesswork had to work with another kind of work: the political and institutional work of renegotiating forms of authority and access to knowledge.

With its feature of parts and wholes, for example, system could be cast as a democratizing vehicle: by explaining many things according to a single principle, it could be presented as a tool for reducing complexity to simplicity. Even typographical conventions played a role. Innovations in the printing of the early Ramist systems, as David Simpson has argued, helped to open writing and knowledge to the "common people" by dispersing the univocal authority of earlier texts: commas and italics set off illustrative materials as coming from a variety of sources, and tables of contents framed the entire text as something not given but made (Simpson 1993, 24–25).

Systems could also, of course, be written and printed to dampen this leveling effect. When skills (and not just ideas) became systemic principles—as in the turn to observation and experiment—older hierarchies of authority were transformed into new hierarchies of expertise: not who has access to knowledge but who knows best how to access it. System making and system reading thus took on increasing social consequence. John Locke, for example, devoted roughly four sections of his *Thoughts on Education* (1692) to the pedagogical role of systems, recurring again and again to the same two-part formulation. On the one hand, the attempt to know "the principles, properties, and operations of things, as they are in themselves" through system was unlikely to succeed:

> though the world be full of systems of it [natural philosophy], yet I cannot say, I know any one which can be taught a young man as a science, wherein he may be sure to find truth and certainty, which is what all sciences give an expectation of. (Locke 1693, 229–230)

On the other hand, systems were not to be dismissed: "I do not hence conclude, that none of them are to be read." For Locke, even flawed systems carried social value: "It is necessary for a gentleman in this learned age to look into some of them, to fit himself for conversation."

Locke's "look," as with Bacon's, engaged system as a genre interrelated with other genres. In his case, however, those others were not forms for making knowledge but pedagogical alternatives. To teach about "spirit," for example, Locke suggested writing a "history" of the Bible complemented by what was then a sister genre of system, the "epitome," "containing the chief and most material heads" (1693, 227). Locke thus relegated system to the status of a schoolboy's shortcut or a minor form of what we now call cultural capital for young gentlemen—with one exception. It was an exception, however, that proved to be exceptional, for it was so widely shared—and shared with such enthusiasm—that it altered the fate of system. That exception was Newton.

Although the *Principia* had been published just six years earlier (1687) than his *Thoughts*, Locke was already writing to the converted when he singled out that book as having changed mankind's understanding of what could be understood. Only when he turned to Newton did Locke even entertain the possibility that knowledge could assume and maintain the form of a system:

> Though the systems of *Physick*, that I have met with, afford little encouragement to look for certainty or science in any treatise … yet the incomparable Mr. *Newton*, has shewn how far mathematicks, applied to some parts of nature, may, upon principles that matter of fact justifie, carry us in the knowledge of some, as I may so call them, particular provinces of the incomprehensible Universe. (232)

Here was the conviction that became so central to Enlightenment—the conviction that the world could be known. What made it comprehensible were new forms of mediation in new combinations. Mathematics made new by Newton's invention of the calculus[4] brought to system a new way of generating "principles" more firmly grounded in "fact"—a regrounding that opened the gap Chambers detected between system and its competitors.

Newton's accomplishment elicited not just admiration but astonishment. The sense that he had changed all of the rules by framing a system of the whole "world" out of a few simple ones was immediate and persistent. But testimonials throughout the eighteenth century were still tinged with surprise that such a thing could be—a "system," as Adam Smith put it at midcentury, "whose parts are all more strictly connected together, than those of any other philosophical hypothesis." Having acquired "the most universal empire that was ever established in philosophy," Newton's principles, asserted Smith, "have a degree of firmness and solidity that we should in vain look for in any other system" (Smith [1795] 1980, 104–105).

But when it came down to specifying what they were actually looking at, admirers and system makers like Smith ran into a really difficult question: how real could a system be? "Even we," wrote Smith, astonished by his own reaction to Newton's system,

> while we have been endeavouring to represent all philosophical systems as mere inventions of the imagination, to connect together the otherwise disjointed and discordant phaenomena of nature, have insensibly been drawn in, to make use of language expressing the connecting principles of this one, as if they were the real chains which Nature makes use of to bind together her several operations. (105)

That "as if" drove the proliferation of system in the eighteenth century, haunting everyone with the possibility that a system could be adequate to the real thing. By portraying that adequacy as a matter of connection, Smith's "as if" echoes the logic of the calculus—of using parts to approximate wholes—as it enacts the scalability of system—of connecting more and more parts into wholes comprehensive enough to "be" real.

This interaction was the engine of Newtonian Enlightenment: *the calculus divided wholes into an infinite number of parts, and system connected parts into wholes.*[5] The *Principia* was the embodiment of that interaction; it literally took shape as a pairing of these tools.

When Newton first took up the project at the urging of Edmond Halley in 1684, he planned a two-book treatise on the "motion of bodies" (*De Motu Corporum*). *Liber primus*, after preliminary matter consisting of "Definitions," a "Scholium," and "Axioms," begins with the principle that undergirds the calculus. "The method of first and ultimate ratios"—of parts approximating the whole—is "for use in demonstrating what follows":

> Quantities, and also ratios of quantities, which in any finite time constantly tend to equality, and which before the end of that time approach so close to one another that their difference is less than any given quantity, become ultimately equal. (Newton 1999 [1687], 433)

Liber secundus then became Newton's first attempt to pair the calculus with the form that makes a whole of parts: with the "motion of bodies" calculated, he proclaimed, the "true constitution of the system would thus be fully and accurately be perceived."[6]

The wider implications for knowledge of pairing of these forms of mediation—*calculating* the world and *systematizing* it—emerged as the project took its final form. By summer 1685, Newton had scaled up his plans. He extended the first book into a new second book on the same topic—the motion of bodies; the old second book was rewritten to become a new third book, now explicitly titled *The System of the World*; and this more comprehensive whole now carried the much more ambitious title of *Philosophiae Naturalis Principia Mathematica*.[7] "Ambition," however, does not really capture what was at stake in this marrying of natural philosophy and mathematics. Newton's new title was, in Edward Grant's (2007) terms, "a virtual contradiction in terms."

The roots of that contradiction lie in Aristotle's classifying natural philosophy, metaphysics, and mathematics as three different kinds of knowledge. The first two were "speculative" rather than "practical" sciences; their purpose was to contemplate truth in regard to changeable bodies—that is, Newton's "bodies in motion." During the fourteenth century, scholastic natural philosophers did begin to apply mathematics to problems of motion, as Grant has shown, but those applications were not understood to be illustrations of "mathematical principles of natural philosophy."

The difference, argues Grant, was that "Newton regarded his treatise as if it were revealing the mathematical structure of physical nature, rather than as the mere application of mathematics to nature, as virtually all previous natural philosophers would have perceived it" (2007, 313). The interplay of the calculus and system, by mixing the mathematically practical and the philosophically speculative, gave shape to a new kind of knowledge. Its unprecedented purchase on the real was signaled by Smith's "as if" and his assertion that Newton's system had "gained the general and complete approbation of mankind" because it was considered

> not as an attempt to connect in the imagination the phaenomena of the Heavens, but as the greatest discovery that ever was made by man, the discovery of an immense chain of the most important and sublime truths, all closely

> connected together, by one capital fact, of the reality of which we have daily experience. (Smith [1795] 1980, 105)

The calculus had provided the mathematical "fact"—the law of gravity—that confirmed the philosophical truth of *system*: that "all" truths were "closely connected together."

Newton acknowledged in a letter to Halley in 1686 that his claim to doing natural philosophy depended on this mix with system. "The first two books without the third," he wrote, "will not so well bear the title of *Philosophiae naturalis Principia Mathematica*." But then the letter zigs and zags in a way that has puzzled many Newton scholars. Having just confirmed that he had been hard at work on the three-book version, he informed Halley of two rapid-fire changes of plan:

> The third I now design to suppress … & therefore I had altered it to this *De motu corporum libri duo*: but upon second thoughts I retain the former title. Twill help the sale of the book which I ought not to diminish now tis yours. (Newton [1686] 1960, 2:437)

A reply from Halley a few days later apparently convinced Newton to return to the three-book plan, thus avoiding a telling discrepancy between the title of the book and its content.

But why did he need convincing? Newton does complain in his letter that

> philosophy is such an impertinently litigious Lady that a man had as good be engaged in Law suits as have to do with her. I found it so formerly & now I no sooner come near to her again but she gives me warning.

Most critics have taken this as a surly reaction to Robert Hooke's demand that he be given credit for the inverse-square law for gravity (Newton 1999, 48). Though that may have been an immediate trigger, "suppression of Book III," as A. Rupert Hall and Marie Boas Hall point out, "could hardly have quietened Hooke, though Newton might have believed that it would do so" (Newton 1978, 234–235).

Whether he did or not, what we do know from this letter is that Newton worried that natural philosophy exposed him and his enterprise to certain risks. The very popularity that promised sales also invited disagreement and controversy—controversy with which we are now so familiar in our age of litigious copyright that we too easily misjudge the scope of Newton's worry. Quarreling over who discovered what and when was certainly a cause for concern for Newton, but what we forget is how *new* a concern it would have been for everyone then. For there even to be squabbles over who made discoveries, knowledge itself had to be understood as something to be discovered rather than reaffirmed. Part of what the Royal Society was acknowledging when it honored Bacon as its father figure was that producing knowledge now meant "advancing" knowledge—and that invited arguments over claiming new ground. It's hard for us to grasp today that the society's motto—*Nullius in verba*, "on the word of no one"—was, back then, the condition of possibility for arguing over who said what first.

Newton may in fact have been hotheaded and disputatious, but focusing on his individual behavior risks our missing how strange—and strained—this newly competitive world of new knowledge must have been for him and for all of his contemporaries. Read with that strangeness in mind, Newton's twists and turns regarding philosophy do not themselves seem so strange. Philosophy had become doubly dangerous for him: at the very moment the tensions of competition heightened, the scholastic version of natural philosophy, as Bacon so vehemently complained, had reached its shelf life. Its toolbox of commentaries and *summas* could not accommodate mathematical principles as Newton wanted to use them.

Newton thus had to make a choice—suppress philosophy *or* change it—that was much tougher and, of course, much more important in the long run than handling Hooke. As the letter makes clear, suppressing was a no-win option. On the one hand, to avoid Hooke and litigation by removing philosophy from the project and reverting to the original title would be to surrender sales and thus the public recognition that was at stake in these rivalries. On the other, the compromise of keeping *philosophy* in the title while suppressing the book that actually contained it would not—to use Newton's word—"bear" even his own scrutiny.

That left changing *philosophy*, and that is where system came in. Newton would have been familiar with system as a vehicle for the scholastic version, a role in which it performed the rhetorical guesswork that Adam Smith derided. That was a cardinal sin for Newton as well; in fact, his problem with Hooke was less about timing and much more about method—about the role of guesswork in producing knowledge. Just like

Johannes Kepler, Newton argued, who "knew" orbits were "not circular but oval," but only "guessed it to be elliptical," so Mr. Hooke

> [knew] that the proportion was duplicate *quam proximè* at great distances from the centre & only guessed it to be so accurately, & guessed amiss in extending that proportion down to the very centre.[8]

Time was clearly not the only variable in these disputes; for Newton, you knew something first only if you *knew* it in the proper way. Guessing was necessary but not enough. In making that distinction, Newton was as tough on himself as he was on others. "I hope I shall not be urged to declare in print," he wrote of his earlier work on the inverse square law, "that I understood not the obvious mathematical conditions of my own hypothesis" (Newton [1686] 1960, 436).

The primary waypoints on Newton's path to understanding—a path trod over and over again in his letters—were "hypothesis," "demonstration," and "philosophy." "Without my Demonstrations," Newton argued, the inverse square hypothesis "cannot be believed by a judicious Philosopher to be any where accurate" (Newton [1686] 1960, 437). Demonstrating the mathematical conditions of earlier guesswork—his own and Hooke's—was the first step to being believed. The next step—from demonstration to philosophy—posed a question that was difficult because it was still new: how do you put a hypothesis cast in the language of mathematics *into* philosophy?

Galileo had most famously put this issue on the table only sixty years earlier with his admonition in *The Assayer* (1623):

> Philosophy is written in this all-encompassing book that is constantly open before our eyes, that is the universe; but it cannot be understood unless one first learns to understand the language and knows the characters in which it is written. It is written in mathematical language, and its characters are triangles, circles, and other geometrical figures; without these it is humanly impossible to understand a word of it, and one wanders around pointlessly in a dark labyrinth. (Galileo [1623] 2008, 183)

Because mathematics did become the language of modern science, celebrations of Galileo's foresight rarely attend to who and what he was admonishing. The who was Orazio Grassi, who published *The Astronomical and Philosophical Balance* in 1619 under the pseudonym of

Lothario Sarsi. But as with Newton and Hooke, it was the what that raised Galileo's ire—and that what was philosophy itself:

> I seem to detect in Sarsi the firm belief that in philosophizing one must rely upon the opinions of some famous author, so that if our mind does not marry the thinking of someone else, it remains altogether sterile and fruitless. (183)

What Galileo detected was the persistence of a mode of knowing into the very medium that would eventually silence it. That mode was the aural one that Ong has eulogized as the sociable discourse of persons, of knowledge arising out of dialogue. To Galileo, this was nostalgic and a mistake. He mocks Sarsi for holding to the notion "that philosophy is the creation of a man, a book like the *Iliad* or *Orlando Furioso*, in which the least important thing is whether what is written in them is true."

"Mr. Sarsi," pronounced Galileo, with a finality intended to cut any further dialogue short, "that is not the way it is" (183). The time had come for books of another kind, books that could, to use Ong's word for knowledge-in-print, "contain" truth because they would be translations of the book that *is* the universe. Galileo's insistence on the language of mathematics as the medium for translation was thus an argument for changing philosophy itself. Philosophy was thus in play in a special way during the seventeenth century: the question of doing it was a matter of debates over the very nature of that enterprise and not just its content. Or, wary of the animosity and pitfalls of those debates, one could decide not to do philosophy at all. As strange as that sounds, that is precisely the choice that ignited the animosity between Hooke and Newton.

Seven years before Newton's letter to Halley, Hooke had written to Newton (November 24, 1679) requesting that he continue his "former favours to the Society by communicating what shall occur to you that is philosophical" (Newton [1679] 1960, 297). This pairing of "philosophy" and "communication" is what connects Galileo's concern about Sarsi's "marrying" of minds (philosophy *as* communication rather than truth) and Newton's decisions about what became the *Principia* (should he communicate truth *through* philosophy?). For Newton, this was a very pressing issue. Moving from demonstrated hypotheses to philosophy meant engaging in acts of communication: how do you put what you want to be believed in writing? If mathematics was the language of demonstration, then must it also be, per Galileo, the language of philosophy, and, if so, then what form should philosophy on mathematical principles take?

In the exchange that followed Hooke's request, this pairing of philosophy with communication surfaces with astonishing regularity. In fact, Newton rarely separates the two, either turning philosophy into an adjective paired with a form of communication or presenting himself as communicating (or not) with philosophy personified. He is "tempted," he writes Hooke, to engage in "Philosophical correspondence," but has "had no time to entertein Philosophical meditations." In that same paragraph, he also describes himself as "endeavouring to bend myself from Philosophy" and then as "having thus shook hands with Philosophy." The letter ends with him "declin[ing] Philosophical commerce," a retreat from communication that provokes a rebuke from Hooke: "Your deserting Philosophy … Seems a little Unkind" (Newton [1679] 1960, 300–304).

Newton and philosophy, in other words, had a history well before their dust-up in the writing of the *Principia*. And his threat of desertion in the letter to Halley is less surprising once we recognize that, by his own admission, he had done so before. It's easy, of course, to psychologize that tale, but I have recovered it not to establish a pattern of personal behavior but to document the fate of philosophy in the late seventeenth century.[9] In the history of mediation, this sequence is not about Newton changing his mind but about changes in philosophy. Deserting philosophy as it had been practiced was the easy part—so easy that Newton repeated the gesture. But to change it into something that he could practice required a retooling—changes that would make it adequate to Galileo's admonition.

From that perspective, the calculus was a new dialect of the language of mathematics, developed to demonstrate hypotheses. For philosophy to communicate what could now be "believed," it too needed a new tool—a tool capable of translating that dialect. This was more reinvention than invention, since Newton chose the same tool—system—that Galileo had helped bring to prominence because it was so helpful to him: system did double duty, both naming what was seen in the physical world *and* turning it into a message. That was precisely the purpose of Book 3 described in Newton's Preface to the first edition. It would, he promised, link philosophy and communication: "Our explanation of the system of the world *illustrates* these propositions" (Newton [1687] 1999, 382, emphasis mine). Suppressing or including that book (the Preface was written in May, a month before the Halley letter) was thus a debate about genre—about whether system could emerge from scholastic inquiry and from the "mere romance" of hypothesis to perform a different kind of work.

This was the intellectual drama behind the often-told melodrama of Newton's spats and also the historical prelude to the pairing I highlighted earlier: the calculus translated wholes into parts, and system translated parts into wholes. Calibrating their interaction, however, was no easy task. Newton did, in fact, suppress his first attempt to "exhibit the system of the world from these [mathematical] principles," revealing in the opening paragraph of the published version[10] that he had

> composed an earlier version of book 3 in popular form, so that it might be more widely read. But those who have not sufficiently grasped the principles set down here will certainly not perceive the force of the conclusions, nor will they lay aside the preconceptions to which they have become accustomed over many years; and therefore, to avoid lengthy disputations, I have translated the substance of the earlier version into propositions in a mathematical style, so that they may be read only by those who have first mastered the principles.

This recalibration put system on what we saw Chambers call "firmer ground." Perhaps appropriately, then, NASA chose the "popular" version Newton rejected to be launched into space. For Newton's immediate purposes, however, mixing in more of the mathematical distanced system from hypothesis, turning it into a different kind of guesswork—a genre more finely tuned to the communicative function of philosophy.

For Newton that function was not simply a matter of clarity and popularity. Part of making his newly demonstrated knowledge believable was to insulate it from knowledge that would erode that belief through "lengthy disputations." That meant stopping the older kinds of knowledge that were generated from such disputation. Newton's strange publication record sets his turn from those kinds and his formal drive to a firmer system in dramatic relief. Newton first published on optics in a 1672 article in the *Philosophical Transactions of the Royal Society*, an effort that elicited a wide range of criticisms requiring detailed rebuttal. For Newton, that kind of debate was not healthy, for it left the knowledge he produced looking like Sarsi's old knowledge—the unconvincing result of deductive hypothesizing and scholastic debate. His reaction was absolute. "Newton was never again," as Charles Bazerman (1988) points out,

> to publish optical results in a journal, nor was he to publish anything else in the *Transactions* or any other journal, except for a minor piece in 1701 on a scale of temperatures. He was to present his major physical findings only within the complete and com-prehensive argumentative systems of the *Opticks* and the *Principia*. (119)

"I thought that publication should be put off," wrote Newton in the Preface to the latter, in order to "publish all my results together" (Newton [1687] 1999, 383). The authority of his two inclusive books lay in their communicating through systems that, in Bazerman's words, "reduced opposing arguments" from debatable differences "to error" (83).

Error correction—once displaced from the public and personal into content and truth—emerged within the books themselves as the process of positing better explanations described by Deutsch. Editions then index progress. Newton justifies the *Principia's* third edition, for example, by emphasizing that "some things ... are explained a little more fully than previously, and new experiments are added," and another "argument ... is presented a little more fully, and new observations" also "added" (Newton [1725/1726] 1999, 400). More things and better guesses alternate, in the manner described earlier, to produce tighter-fitting explanations.

System assumed a central role in this reshaping of knowledge thanks in part to what Newton did to it. Not only did he put it onto firmer ground by altering its "style," but by choosing it as the form for consolidating and conveying what his new principles could demonstrate, he provided the "as if" exception to system's reputation for falling short. Its growing stature was indexed by how people referred to the *Principia*. The dedication calls it a "treatise most humbly" presented to James II—a genre, as we have seen, commonly employed for knowledge work during the previous century and often a home for systems. But Book 3's "System of the World," through the *pars pro toto* (part for whole) of synecdoche—system's sister among figures of speech—came to stand for the entire work. The *Principia* became known as the "Newtonian System."

MONSTERS OF MEDIATION—THE SUBLIMITY OF SIMPLICITY

In its extensive use of maths, the *Principia* was highly complex—its details decipherable only by a very few. But experienced synecdochically as a system, it conveyed an unprecedented sense of simplicity. What Locke admired as exceptional about Newton's system was not only its "as if" quality but its clarity—and what that promised for similar endeavors:

> And if others could give us so good and clear an account of other parts of *Nature*, as he has of this our planetary world, and the most considerable *Phoenomena* observable in it, in his admirable book, *Philosophiae naturalis principia Mathematica*, we might in time hope to be furnished with more true and certain knowledge in several parts of this stupendious machin, than hitherto we could have expected. (1693, 232)

Once Newton had demonstrated that system could be made "good and clear," it became a primary genre for negotiating the relationship between complexity and simplicity in other parts of the machine of Nature.

That negotiation became a crucial form of Enlightenment mediation. Its project of "more true and certain knowledge" of more and more "parts" took the inverse square law as its model. The promise of the *Principia* was that everything could be known because it could be known through such laws. And the premise behind the "could" was that both Nature and its laws shared one fundamental characteristic: simplicity. Since laws reduce the apparently complex to the simple, then the things they claim to describe must themselves be simple.[11] Simplicity is thus Newton's starting point for the "System of the World." The first of the "Rules for the Study of Natural Philosophy" establishes it as the condition of possibility for the entire enterprise:

> Rule 1 *No more causes of natural things should be admitted than are both true and sufficient to explain their phenomena.*
>
> As the philosophers say: Nature does nothing in vain, and more causes are in vain when fewer suffice. For nature is simple and does not indulge in the luxury of superfluous causes. (Newton [1687] 1999, 794)

This Ockhamite proclamation, in its insistence on sticking to the sufficient, cuts off from consideration anything that is more than enough. Knowledge was thus reclassified as well as produced. Metaphysics, for example, once contiguous with natural philosophy as a speculative science, was now ruled out of the system. Like the supernatural, the metaphysical was an example of "more," and thus placed beyond the bounds of the causal.

Boundaries, like parts and wholes, have been, as noted earlier, a persistent feature of the genre of system, playing a more or less prominent role in different deployments of the genre. It became particularly prominent in specific fields of knowledge making, such as thermodynamics, where a system is explicitly understood as the part of an environment under consideration. The effect of any "working substance," such as steam, can then be consistently measured within the set confines of the system. In the twentieth century, this constituting of systems through boundary setting became a basic experimental protocol of the modern practices of system modeling and systems theory.

By starting his System with "Rules," Newton helped to establish that protocol for knowledge making. Rule 1 defined Nature as a simple system, and rule 2, as we saw in chapter 1, insisted on that system's internal scalability. If apples behaved like moons for the same simple reason, then gravity could be expressed as a "law" governing both. That law, in turn, materialized the notion of a "solar system" as a physical entity: all things that are under the influence of the gravitational pull of the sun. Composed and used in this manner, system thus embodied Bacon's handshake of the intellectual and visible worlds: system became a thing in the world *and* a way of constituting that world as a thing.

The "Newtonian Moment," to use Mordechai Feingold's (2004) phrase for the immediate and astonishing impact of Newton and his work in the late seventeenth and eighteenth centuries, was thus a major event in the history of mediation. A new process of knowledge making was encoded within a specific genre: the Newtonian system demonstrated what could be gained by setting boundaries and exploring the relationship between parts and whole within those boundaries. Although systems did continue to be blamed for failing to explain and for producing "imaginary" philosophers, the *Principia* was a watershed in the attitudes toward and the use of system. Newton was widely understood to have provided an example of what system could do when, in Chambers's words, it was not only "built on the firmest Ground" but also "raised by the severest Rules."

As Newton's system convinced others, in Britain and across Europe, that the entire world really was a system, the genre's logic of scale and simplicity fueled its proliferation. Almost a century after the *Principia* was published, the abbé de Condillac still celebrated Newton's achievement as a matter of system and its consequences:

> Since the universe can be simplified as a system, each part of it having the least complexity is a system: man himself is a system. If, then, we renounce systems, how can we explore anything deeply? I agree that in general philosophers are wrong. They invent systems, but systems should not be in-vented. We should discover those which the author of nature has made. (Condillac [1749] 1947, 3.511–3.512)[12]

Writers of systems, that is, needed to be good readers—of both the divine author and the deductive "errors that the craze for systems led to" (Condillac [1749] 1982, 10). The former told of man's place in nature as part of things as they are, while the latter detailed his departures from it when pressed to explain those things.

In pairing the divine and system, Condillac had the precedent of Newton's turn to God in the "General Scholium." Added to the end of the *Principia*'s second edition (1713) as a comment on the main books, it begins by contrasting the "hypothesis of vortices" with the explanation of bodies in motion that Newton had just demonstrated. By tagging Descartes' effort as a "hypothesis" at the start of the "Scholium," Newton set the stage for ending it with the catchphrase that became the motto of Newtonian science: "*Hypothesis non fingo*" ("I do not feign hypotheses"). In the paragraphs between, Newton gave his alternative form of guesswork—guesswork modified (in the mathematical "style") to make it demonstrable—a divine seal of approval:

> This most elegant *system* of the sun, planets, and comets could not have arisen without the design and dominion of an intelligent and powerful being. (Newton [1713] 1999, 940, emphasis mine)

With that system as the template, the rest of the universe falls simply into place:

> And if the fixed stars are the centers of similar systems, they will all be constructed according to a similar design and subject to the dominion of *One*, especially since the light of the fixed stars is of the same nature as the light of the sun, and all the systems send light into all the others.

All of the other systems will be just like the one system Newton described in Book 3—the oneness of all systems evidence of the dominion of the *One* God.

We need that evidence, Newton insisted, because "without dominion," God "is not the Lord God." We can thus "know him" only through our knowledge of his dominion: "only by his properties and attributes and by the wisest and best construction of things and their final causes." After the hesitation and revision I have documented, Newton chose system as the tool that could best construct things with the wisdom to demonstrate

> that gravity really exists and acts according to the laws that we have set forth and is sufficient to explain all the motions of the heavenly bodies and of our sea.

That effort at causation—per rule 1, stop when you have a true and sufficient one—was the closest we could get, claimed Newton, to final causes, "for all discourse about God is derived through a certain similitude from things human, which while not perfect is nevertheless, a similitude of some kind." "To treat of God from phenomena," he concluded, "is certainly a part of 'natural' philosophy" (941–943).

Instead of excluding God from or installing Him into this newly capacious version of natural philosophy, Newton made "godhood"[13] an optional topic—a part that could be included (as in the *Principia*'s second and third editions), or not (as in its first), by those who did or did not share his belief. Either way, the exact same phenomena remained the actual objects of study. As the tool of choice for constructing this newly popular kind of knowledge, system's popularity rose—and, in treating phenomena in a simple way—system then repaid the favor, enhancing natural philosophy's broad appeal. "A system is nothing other," argued Condillac,

> than the arrangement of different parts of an art or science in an order in which they all lend each other support and in which the last ones are explained by the first ones. Parts that explain other parts are called principles, and the fewer principles a system has the more perfect it is. It is even desirable to reduce all principles to a single one. (Condillac [1749] 1982, 1)

What systems want, in other words, is simplicity—in this case, the highest ratio of parts to principles. The most desirable and most common method through most of the eighteenth century, as recommended here, had been an emphasis on fewer and more comprehensive principles—for example, one law to explain motions of all kinds. Toward the end of the century, however, with more—and more specialized—systems, ratios were raised by increasing the diversity and thus number of parts.

Such simplicity could, of course, be seen as reductive rather than productive, and accusations of that kind constituted another chapter in system's history of blame. In 1759, for example, Oliver Goldsmith wielded that complaint as a weapon in the increasingly heated rivalry between Britain and France. His chapter on the French in *The Present State of Polite Learning in Europe* begins by complimenting them on a "growth of genius … more vigorous than ours." He credits that in part to "the fair sex in France" having "not a little contributed to prevent the decline of taste and literature, by expecting such qualifications in their admirers."

Unlike the "damsels" in Holland who can be "caught"

> by dumb shew, by a squeeze of the hand, or the ogling of a broad eye, [French women] must be pursued at once through all the labyrinths of the Newtonian System. (Goldsmith 1759, 100, 102–103)

Navigating those labyrinths was, I have described, a matter of negotiating the relationship between complexity and simplicity. Goldsmith was not critical of the former—handling complexity kept genius on its toes—but the latter, in his analysis, had become a stumbling block for the French:

> The writers of this country have of late also fallen into a method of considering every part of art and science, as arising from simple principles. … To this end they turn to our view that side of the subject which contributes to support their hypothesis, while the objections are generally passed over in silence. Thus an universal system rises from a partial representation of the question, an Whole is concluded from a Part. (112–113)

Bearing witness to system's pervasive presence by midcentury, Goldsmith claims to find those risings and conclusion in "almost every subject," even those "naturally proceeding on many principles" are "all taught to proceed along the line of systematic simplicity" (112–113).

Goldsmith's critique is particularly valuable for our understanding of system's proliferation because he recognized that this line did not dead-end in simplicity. Thanks to system's scalability, a simple whole could become a part of a more complex one. Just as the labyrinthine had been reduced to the simple, so the simple could scale up to its apparent opposite. In Goldsmith's rendering, the line of simplicity scaled in both directions: either "contracting a single science into system" or "drawing up a system of all the sciences united."

Goldsmith's description of the latter evokes for us now Mary Shelley's bad dream of a half-century later. "Such undertakings as these," Goldsmith warned,

> are carried on by different writers cemented into one body, and concurring in the same design, by the mediation of a bookseller. From these auspicious combinations, proceed those monsters of learning, the Trevoux,[14] Encyclopedie's, and Bibliotheques of the age. (114)

These monsters of mediation were large scale in every way. Most obviously, they dwarfed in physical size the pioneering encyclopedic efforts from the previous century, such as Alsted's—leaving readers "daunted at the immense distance between one great pasteboard and the other" (115). But they were also immense undertakings in human terms. Thanks to Newton's example and popularity, these systems were not single-authored efforts aimed at a small circle of scholars, but "in making these, men of every rank in literature are employed, wits and dunces contribute their share" to the "enormous mass" that "makes its way among the public" (114).

At this scale of pages, writers, and readers, system evoked the sublimity of simplicity. He who enters the "wide extended desart" of these volumes, warned Goldsmith, discovers that "what is past only encreases his terror of what is to come" (115–116). In criticizing the French, he does highlight the negative effects of simplicity and scaling—its "terror" and the "perplexity" it produces—but the admirable and powerful facets of sublime aspirations surface in the use of words like "great" and "genius."

MIXED MESSAGES—"TOWARDS" SYSTEM

This mixed message of affirmation and critique—Newtonian optimism regarding what could be known and the record of ambition gone awry—is a characteristic feature of mid- and late eighteenth-century attitudes toward system and of the systems themselves.[15] A letter to the *Gentleman's Magazine* in 1777, for example, began with an obligatory bow to Bacon and then proceeded to turn a discussion about inoculating the poor into a debate over system: when men's

> opinions are warped in favour of a System, all future experiments must be made to fit it. Thus from some successful experience of the benefit of tar-water, Bp. Berkeley erected a fanciful and elaborate theory, which attributed to it the essence of all medical virtues. ("Letter," 1777, 105)

A writer in the *Monthly Review*, however, had earlier cited precisely the same figure not to attack but to extol the "spirit of system":

> To the discerning enquirer after philosophy and science, the speculations of a Berkeley or a Hume, notwithstanding the absurdities with which they may be chargeable, are infinitely more valuable than the collective mass of the dissertations and essays that have been written against them. ("Review of Essays Moral, Philosophical, and Political," 1772, 382–382)

Notice how a disagreement that appears to be about the work of an individual turns out to be about the form of that work: the systematic speculations of Berkeley versus the essays of his critics.

David Hume's career famously enacted that debate. When his systematic *Treatise of Human Nature* (1739–1740) "fell dead-born from the press," he switched genres, fragmenting it into separate inquiries into "understanding" (1748), "morals" (1751) "passions" (1757), and "religion" (1755), and then arranged to have them published together posthumously as *Essays and Treatises on Several Subjects* (1777). The status of "system" played a crucial role in the content as well as the form of those efforts. He used system both to explain how skepticism works—focusing on the conditions under which we should "embrace a new system with regard to the evidence of our senses"—and to clarify its purpose. Skepticism, he argued, is not a turn from the production of knowledge but a "necessary preparative" to do it well: "though by these means we shall make both a slow and a short progress in our systems; [they] are the only methods, by which we can ever hope to reach truth" (Hume [1758/1777] 1902/1975, sec. 12, 150–152).

Thomas Pownall's *A Treatise on the Study of Antiquities* (1782) is one of the most telling examples of this debate over system, for rather than indulging in abstract praise or satire of others, Pownall posed to himself the practical problem of finding the best genre for a particular kind of work. For studying antiquities, he argued, the choice comes down to making systems or collections:

> Did we follow the seductions of fancy, and quitting the sober steps of experience, hastily adopt system; and then form a dotage on our own phantoms, dress such system out in the rags and remnants of antiquity, we should only make work to mock ourselves: or were we on the other hand to persevere in making unmeaning endless collections without scope or view, we should be the dupes of our futility, and become in either case ridiculous. (Pownall 1782, 3)

The risks of both genres are cast as risks to the health of those who use them: "The upstart fungus of system is poison to the mind; and an unintrusive mass of learning may create and indulge a false appetite, but never can feed the mind … [for] all the learning in the world, if it stops short and rests on particulars, never will become knowledge." Pownall believed that his colleagues—tempted by "extreams of self-delusion on one hand, or of the false conceptions of barren folly on the other"—could choose the right genre only by keeping their "minds constantly fixed on the PRINCIPLE and END of our institution" (4).

That institution was the Society of Antiquaries, and what Pownall termed its "branch of learning" did require, despite the risks, a system—but a system of a particular kind and constructed in a particular way. As antiquaries, Pownall argues, their study must be the study of "the system of the human being; and of the state of nature, of which that being is a part." Since that state changes over time, the antiquary must trace the "great machine" of that system "in all its parts" through "every period of its progressive existence, and compare all with the present state of it" (4). Pownall's strategy for antiquarianism thus turns out to be the "gratifying" relationship between system and history that I described using the tectonic maps of chapter 2. As a product of the last two decades of the century, it corresponds chronologically to the third of those maps in which system and history came to share a tectonic edge. To produce a historical system, Pownall advises, the analysis "must" avoid the "theoretic abstract view of things in general" that leads to "self-delusion" and follow instead "the path in which nature *acting* leads; and by a strict induction of her laws as found in her actions" (5).

Pownall understood that path to nature's laws through system to be Newton's path. System under the rule of simplicity is not a fungus, nor can it be extreme. In one of the most clear-cut examples of the shape knowledge takes from the Enlightenment, Pownall pairs that Newtonian "line" with the "concurrent" one of Bacon and his tools: "what the lord Verulam calls "*Inventarium opum humanarum*"—the "inventory of human resources" (5–6). That line of mediation should be "traced," argued Pownall, using Bacon's strategy of multiple histories, headed by the "experimental history" that Bacon posited as a primary vehicle for "Reinstauration." As Bacon's line joins Newton's, the renewal of knowledge takes, for Pownall, the form of a system. It does so because the Newtonian exception—the "as if" of his particular system—convinced him, like so many others, that "there is *in* nature a system" (emphasis mine). Using Bacon's historical tools, Pownall aspired to translate the system of nature into a knowledge system of linked "causes and effects" that "investigation may retrace back" to the "principles on which the whole depends" (6–7).

In Pownall's case, then, the mixed message of critique and affirmation of system set the stage for treating system not just as a better genre, but as the genre that nature's own system calls on us to reproduce. Whatever the failings of individual systems—in fact, because they could and did have failings—system became a primary form of the Enlightenment. In addition to its empowerment by history, we can gauge that ascension through system's relations with other forms, such as essay, and with particular types of organization, such as alphabetization.

The competition between system and essay throughout the eighteenth century—documented, as we have seen, by Johnson in his *Dictionary* and by my title page counts in figure 1.5—emerges as well in the titles themselves and in comments that invoke both genres. One phrase in particular was repeatedly used to suggest the direction knowledge was now supposed to take in response to nature's call to reproduce its own system on the page. Right at midcentury, for example, John Burton sought to recover his reputation as a physician after his arrest for treason for Jacobite leanings after the rebellion of1745. His primary tactic was to publish a work on obstetrics to compete with William Smellie's *Theory and Practice of Midwifery*. The strategy was twofold: include engravings (the first illustrations published by George Stubbs, who later gained fame for his paintings of horses) and set the generic bar as high as possible. His title turned Smellie's nouns into adjectives modifying his own project's lofty ambition: *An Essay Toward a Complete New System of Midwifry, Theoretical and Practical* (Burton and Stubbs 1751).[16]

"Essay towards system" conveyed the Newtonian aspiration of Enlightenment—the sense of being generically adequate to the real. System became the preferred form of attention to nature. By 1771 Horace Walpole could put this generic preference to what we would now call aesthetic ends with this assessment of the landscape architect William Kent in "The Modern Taste of Gardening":

> When nature was taken into the plan, under improvements, every step that was made, pointed out new beauties and inspired new ideas. At that moment appeared Kent, painter enough to taste the charms of landscape, bold and opinionative enough to dare and to dictate, and born with a genius to strike out a great system from the twilight of imperfect essays.[17]

Other techniques for striking out systems from essays were not quite as daring. Perhaps the most mundane was alphabetization. Adding more was one way to improve on the imperfect, and the alphabet facilitated that process by providing a template for organizing more parts into a limitless whole without requiring new categories.

The alphabet thus occupied the borderland between collection and system that so concerned Pownall. In fact, Johnson's definition of the genre that was most closely identified with the alphabetical—the dictionary—cited Isaac Watt's description of that genre as "a collection of words." But such collections also became literary Petri dishes for growing new generic strains by modifying the alphabetical code. Diderot and d'Alembert, for example, as Richard Yeo has shown, praised Chambers "for his attempt to impart system to an alphabetical dictionary" by adding cross-references to his *Cyclopaedia (or Universal Dictionary)*—such as those we followed earlier between "system" and "hypothesis" (Yeo 2001, 26).

The French efforts "toward" system in the *Encyclopédie*, however, were not enough to prevent William Smellie (another Smellie, not Burton's rival) from turning them into a selling point for the 1771 first edition of his *Encyclopaedia Britannica*:

> Whoever has had occasion to consult Chambers, Owen, etc. or even the voluminous French *Encyclopédie*, will have discovered the folly of attempting to communicate science under the various technical terms arranged in alphabetical order. Such an attempt is repugnant to the very idea of science, which is a connected series of conclusions deduced from self-evident or previously discovered principles.

For Smellie, knowledge is shaped by the relationship between "parts" and wholes—a relationship governed by Newtonian "principles" and thus incompatible with the alphabetical. "But where is the man," asks Smellie, "who can learn the principles of any science from a Dictionary compiled upon the plan hitherto adopted" (*Britannica* 1771, Preface 1).

Smellie announced his "NEW PLAN"—one that would allow "any man of ordinary parts" to learn principles—on the title page of the first edition and then dwelled on the intention of the "compilers" in the Preface:

> Instead of dismembering the Sciences, by attempting to treat them intelligibly under a multitude of technical terms, they have digested the principles of every science in the form of systems or distinct treatises, and explained the terms as they occur in the order of the alphabet, with references to the sciences to which they belong.

The editors of *Britannica*'s third edition celebrated their predecessors' initial modification of the alphabetical as a necessary switch of genre. They understood Smellie to have turned to "system" as the generic alternative to "collection"—the competition's "vast collection of particular truths" that was of "little advantage to the arts of life." To turn to system was, per Newton, to turn to nature itself: "Much cautious attention is requisite to class objects in human systems as they are *in fact* classed in the system of nature" (*Britannica* 1797, Preface, 1–2, emphasis mine).

By that third edition, in the kind of change we now call "emergence" (Siskin 2005, 819), more and more classification, in *Britannica* and in many other venues, yielded difference—a difference in the organization of knowledge itself that I explore in the next chapter. Smellie's *Encyclopaedia* was not even the only "Britannica" to feature this turn toward system. Compare the two title pages from 1747 and 1769 in figure 3.4.

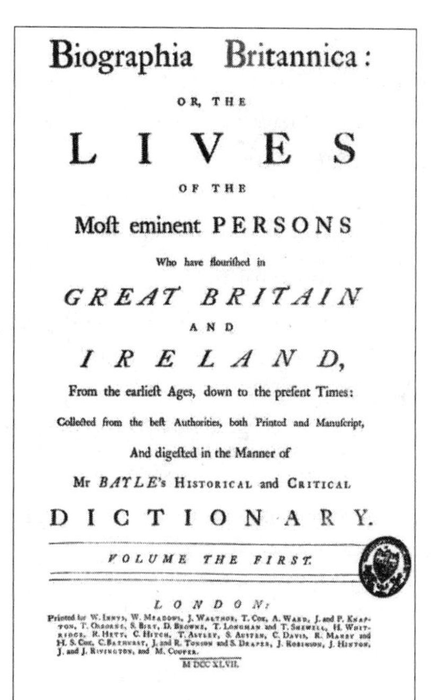

Biographia Britannica:

OR, THE

LIVES

OF THE

Moft eminent PERSONS

Who have flourifhed in

GREAT BRITAIN

AND

IRELAND,

From the earlieft Ages, down to the prefent Times:

Collected from the beft Authorities, both Printed and Manufcript,

And digefted in the Manner of

Mr *BAYLE*'s HISTORICAL and CRITICAL

DICTIONARY.

VOLUME THE FIRST.

LONDON:

Printed for W. INNYS, W. MEADOWS, J. WALTHOE, T. COX, A. WARD, J. and P. KNAPTON, T. OSBORNE, S. BIRT, D. BROWNE, T. LONGMAN and T. SHEWELL, H. WHITRIDGE, R. HETT, C. HITCH, T. ASTLEY, S. AUSTEN, C. DAVIS, R. MANBY and H. S. COX, C. BATHURST, J. and R. TONSON and S. DRAPER, J. ROBINSON, J. HINTON, J. and J. RIVINGTON, and M. COOPER.

M DCC XLVII.

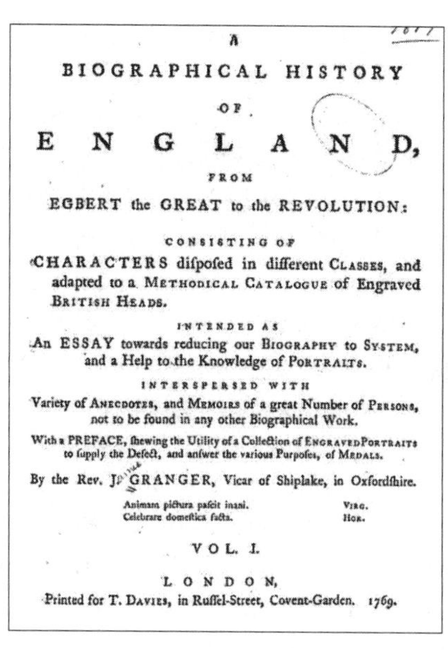

A

BIOGRAPHICAL HISTORY

OF

ENGLAND,

FROM

EGBERT the GREAT to the REVOLUTION:

CONSISTING OF

CHARACTERS difpofed in different CLASSES, and adapted to a METHODICAL CATALOGUE of Engraved BRITISH HEADS.

INTENDED AS

An ESSAY towards reducing our BIOGRAPHY to SYSTEM, and a Help to the Knowledge of PORTRAITS.

INTERSPERSED WITH

Variety of ANECDOTES, and MEMOIRS of a great Number of PERSONS, not to be found in any other Biographical Work.

With a PREFACE, fhewing the Utility of a Collection of ENGRAVED PORTRAITS to fupply the Defect, and anfwer the various Purpofes, of MEDALS.

By the Rev. J. GRANGER, Vicar of Shiplake, in Oxfordfhire.

Animum pictura pafcit inani. VIRG.
Celebrare domeftica facta. HOR.

VOL. I.

LONDON,

Printed for T. DAVIES, in Ruffel-Street, Covent-Garden. 1769.

Figure 3.4
Title pages from 1747 and 1769.

Volume 1 of William Oldys's *Biographica Britannica* announced itself in 1747 as "collected" and then "digested in the Manner of" Bayle's *Dictionary*. Pierre Bayle's *Dictionnaire Historique et Critique* was first published at the end of the previous century (1695–1697) and translated into English in 1709. Its lives were collected into a strictly alphabetical format, but Bayle, like Chambers, sought to surpass his predecessors within the entries themselves; his technique was not cross-referencing but a focus on compilation, comparison, and skeptical critique.

Richard Popkin argues that Bayle's influence on Enlightenment figures such as Oldys—figures ranging from Voltaire to Hume—entailed a misunderstanding. "Bayle's message," Popkin asserts, "was that of the inadequacy and incoherence of man's intellectual endeavors, of the need for a different guide—faith and revelation." But that stance was "transformed," Popkin argues, "by the Age of Reason into a positive affirmative of other views, into a new theology—scientism" (Popkin 2003, 300). The notion of "thinkers" misunderstanding each other fits well with the "Ages" and "-isms" of the history of ideas, but does not help us explain what happened and why. Within a history of mediation, however, we can engage Bayle's effort as participating in the remediation of scholastic forms, including scholastic systems. Like Newton, his contemporary, but in a different way, Bayle's skeptical dismantling of established kinds of knowledge cleared the way for the reshaping I am describing; his work, that is, was not mysteriously misunderstood but enabling.

James Granger launched his challenge to Bayle's and Oldys's "manner" of proceeding within one year of Smellie's new plan for encyclopedias. The 1769 title page highlights the primary strategy common to both endeavors: these were turns to system. Granger actually used the standard phrasing, describing his project as an "ESSAY towards ... SYSTEM." And since he called the volume a *Biographical History*, it is yet another example of system and history sharing the same late eighteenth-century title page. Their interrelations shaped Granger's effort to—as Smellie's successors put it—"class objects."

In a lengthy appraisal, the *Monthly Review* focused on that effort, providing an extended extract of "The Author's plan." This sample highlighted the interaction of historical classification—

> All portraits of such persons as flourished before the end of the reign of Henry the Seventy, are thrown into one article

—with other kinds:

> In the succeeding reigns, they are ranged in the following order:
>
> Class I. Kings, queens, princes, princesses, &c. of the royal family.
> Class II. Great officers of state, and of the household. ("Review of Granger's Biographical History of England," 1769, 209)

Many more classes follow, illustrating once again system's role in negotiating the relationship between the simple and the complex. Although Granger described the move toward system as an act of "reducing," the reduction took the form of dividing and including; the drive to simplify by grouping things into different classes, that is, heightened attention to difference—thus inviting more classes.

That invitation was a major reason that Granger, as Felicity Nussbaum notes, was "among the innovative few that reflect on the method for representing female character in biography" (2005, 308). In addition to the initial chronological grouping, Granger topped out at twelve different classes, including not just "Ladies, and others, of the female sex" but also others

> of both sexes, chiefly of the lowest order of the people, remarkable from only one circumstance in their lives; namely, such as lived to a great age, deformed persons, convicts, &c. (Granger 1769, 2–3)

The move toward system thus changed what was being systematized, transforming the project of national biography from Oldys's collection of the "Most eminent PERSONS" into an attempt at ("essay") a new whole consisting "of CHARACTERS disposed in different Classes."

This apotheosis of system in the ambitions of Smellie, Granger, and many others was furthered by system's interrelations with other forms. History, as I am documenting, played an important role here, but system's new claims and status were most clearly and grandly enacted through a form of self-improvement: the embedding of systems within Systems. For a brief period of time in the mid-and late eighteenth century, the decades we most commonly brand Enlightenment in the West—the genre that was used to trump system was system. Here, for example, is Benjamin Martin in 1747:

> Having read and consider'd the Design of the several Books hitherto published for the Explanation of the NEWTONIAN PHILOSOPHY, under the titles of *Commentaries, Courses, Essays, Elements, Systems*, &c. I observed not one of them all could be justly esteemed a TRUE SYSTEM, or COMPLEAT BODY of this science. (1)

"SYSTEM" rises from this long list of genres to be singled out as the "TRUE" form of "science" here in Martin's preface and in his title: *Philosophia Britannica or A New and Comprehensive SYSTEM of the Newtonian PHILOSOPHY, ASTRONOMY and GEOGRAPHY.* The solution to failed, partial, or old systems and other forms of generic inadequacy was system itself—a Master System ambitious enough to comprehend the embedded remains of its predecessors.

MAKING MASTER SYSTEMS—STRATEGIES FOR ENLIGHTENMENT

Using the upper case for this type of system is more than a matter of emphasis; it conveys more specifically that Master Systems were the product of one of the genre's characteristic features: its capacity to scale up. The particular ambitions of Master Systems varied, as did their modes of comprehensiveness, but making them always entailed the manipulations of parts and wholes. Benjamin Martin's ambition for his System was to compose a "*General Compendium* or *Abridgment*" of the "Abundance of Materials supplied by [his] Authors." His mode, like Granger's, was to "reduce," and since each author was "excellently good in his Way," Martin opted for "select[ing] the best of every Thing I could find for my own Composition" (1747, Preface, 1–2).

The remediation of system by System was, I am arguing, a central activity of the Enlightenment project to know the world, particularly as it was pursued in Scotland as part of the British Enlightenment. By "project" I mean that Adam Smith and his Scottish cohort did not just stumble into the making of Master Systems. Far from being just a happy meeting of great minds, a fortuitous flowering of ideas, the Scottish endeavor was a carefully planned foray into the newly expanding world of print, what today we might call—minus the cynicism—a public relations ploy. Writing in 1755 in the *Edinburgh Review*, Smith surveyed the scene and laid out the plan. At that historical moment, he admitted, Scotland, "which is but just beginning to attempt figuring in the learned world, produces as yet so few works of reputation, that it is scarce possible a paper which criticises upon them chiefly, should interest the public for any considerable time" (Smith [1795] 1980, 242). For Smith, this was no cause for despair but rather an opportunity for Scotland to carve out a special place in that world. He did so by putting the other European countries in their places, grading their achievements in each kind of knowledge.

In natural philosophy, he noted, the stars are England, Germany, and Italy. Hobbled by the Cartesian philosophy for too long, France was just beginning to catch up, if not in its writing, then in its reading: "it is with pleasure that I observe in the new French Encyclopaedia, the ideas of Bacon, Boyle, and Newton" (244–245). The pleasure Smith felt is a window into the complexities of British national identity at midcentury: "As, since the union, we are apt to regard ourselves in some measure as the countrymen of those great men, it flattered my vanity, as a Briton, to observe the superiority of the English philosophy thus acknowledged by their rival nation." But here's the Scottish twist. At the very moment of apparent identification with the English under the rubric of Britishness,

Smith identified a gap between the two categories—English and British—a gap in which something distinctly Scottish could flourish.

In a turn so quick the English probably still don't know what hit them, Smith suddenly describes himself as "mortified" by the very achievements that had flattered him. Casting the "renowned philosophers" he had just named as exceptions, Smith dwells on how many English writers are "unknown" and "disregarded in their own country." He is embarrassed, he claims, by the fact that "posterity and foreign nations are more likely to be made acquainted with the English philosophy by the writings of others, than by those of the English themselves." As we have seen, Newton took system's communicative possibilities only so far, retreating, in the published version of the *System of the World*, from the "popular" to the less accessible "mathematical style." Here, then, was the opening for Scotland:

> The English seem to have employed themselves entirely in inventing, and to have disdained the more inglorious but not less useful labour of arranging and methodizing their discoveries, and of expressing them in the most simple and natural manner. There is not only no tolerable system of natural philosophy in the English language, but there is not even any tolerable system of any part of it. (245)

Scotland, Smith realized, could enter the learned world by writing systems—more specifically Master Systems that would arrange and methodize all earlier systems. That's why Smith's major works always included specific sections for comprehending the competition within a larger whole. And that's why "simple and natural" expression was not a polite afterthought for Smith, Hume, and the other writers of the Scottish Enlightenment, but a guiding first principle of their joint effort to assume a place among the learned nations by, in part, succeeding where they thought the English had failed or lapsed.

With this strategy of Master Systems in hand, no question was out of the question: everything could be known. Thus, a mere four years after targeting moral philosophy as a "branch of the English philosophy, which seems now to be intirely neglected by the English themselves" (250), Smith took as the task of his very first book (1759) an explanation of the principles of human behavior. Of the seven parts of *The Theory of the Moral Sentiments* ("theory," according to Johnson, being a "system yet subsisting only in the mind"), the longest is the final one; there we find embedded the "particular system[s]" formed out of the "different theories" of his predecessors (1976, 265). My point here is not that earlier writers neglected to review their competition—or that other genres did not feast on their own kind—but that system turned on itself in a particularly powerful way. Thanks to its scalability, system became a historical site for extraordinary intellectual aspirations mixed with sustained attention paid to the very genre that articulated them.

Smith's first book ended by identifying another gap in knowledge as another opportunity for system: the lack of a recent complete "account of the general principles of law and government" ([1759] 1976, 342). He filled that gap with what became his most famous Master System, *The Wealth of Nations*. In Smith's hands, as John Gray points out, political economy became distinctively "systematic and comprehensive [in] character" (Gray 1995, 25). "There are no chapters in *The Wealth of Nations*," observes Athol Fitzgibbons, "called 'The optimal Allocation of Resources', or 'The Efficiency of Free trade'; but there are chapters called 'Of Systems of Political Economy', and 'Of the Principle of the Commercial, or Mercantile, System', because Smith wanted to compare his liberal system with other general systems" (Fitzgibbons 1995, 172).

Following his own formula for system—in which one principle is "found to be sufficient to bind together all the discordant phenomena that occur as a whole species of things" (Smith 1980, 66),[18] Smith foregrounded the "division of labour." And within that overarching system, he embedded an entire book (one of five) on the various "Systems of political Oeconomy" in "different ages and nations" (Smith 1776, I.428). But even as he scaled up his own system by incorporating others, the very logic of that scalability compelled him to apologize for not going far enough; he expressed regret for not adding a "theory of jurisprudence" due to "very advanced age." That admission, however, only added to the sense of explanatory power attributed to *The Wealth of Nations*. Its reputation arose in part from the conviction that all previous systems could be productively reconciled in this age and in this nation; it was an effect, that is, of the process of embedding systems within a Master System.

By "effect" I mean to highlight how the sense being communicated is less a matter of content than of form. The role of Master Systems in generating Enlightenment was not a matter of any particular set of *ideas*, but of the particular ways in which they were mediated. To identify Enlightenment with the conviction that the world could be known—known because the formal representations we call knowledge seemed to resemble it so closely—is to highlight the form of that resemblance. That form, as I have been documenting, was system, and the effect of remediating system by System was to change the object to be known, as well as the shape of that knowledge.

To write up the "world" as a system, per Newton and then explain it through more and more encompassing Systems, per Smith, was to shape what that world came to be. We might say that it was prescribed, that is, written before it became real—real to those who then took it *as* a system and thus something that could be known.[19] Knowledge of that world thus took the form of more writing and new institutions, for the effects of enfolding systems within System reverberated into and out of print in multiple and complementary ways. The year 1759, for example, saw not only Smith publishing his first Master System of moral philosophy but also the opening of the British Museum as what its director in 2009 called "the first coherent intellectual response to globalization" (Hoyle 2009). While Smith's purpose was to fit Scotland into the encompassing SYSTEMS of the United Kingdom and the Republic of Letters, the museum was founded to enfold Sir Hans Sloane's bequest of his personal collection into a newly comprehensive organizational System—one that materialized as and in a new kind of container.

At the time of Sloane's will in 1753, the word *museum*, as Laura Yoder points out, was "a direct synonym of the word 'collection,' and refer[red] to the things rather than the building." Only after the doors on Great Russell Street opened did "museum" fully take on its modern meaning (Yoder 2012, 27).[20] That opening enacted the movement from collection to system we saw earlier in Pownall, but now scaled up organizationally and physically through a double enfolding: the personal (Sloane) entered into the national (British) as the national entered into the global. Britain thus assumed its place in a new global Master System of nations that its own new institution helped to institute by accumulating and assimilating more collections.

The institution of the national museum can thus be understood as another formal *effect* of the system-based strategies that shaped knowledge and its objects during the Enlightenment. Identifying those strategies and, crucially, how they changed, is a central task of this book, for what's at stake in my subtitle—*The Shaping of Modern Knowledge*—is not only what we have inherited from Enlightenment but also our distance from it today. In the history of ideas, modern knowledge is often seen as developmentally continuous with Enlightenment, with recurring ideas providing the continuity. But in the history of mediation, a crucial difference becomes visible: modern knowledge emerged from the Enlightenment as something different. That difference, I will argue, was an effect of new deployments of system.

4 DISCIPLINARITY (EMBEDDED AND SPECIALIZED SYSTEMS)

Is not a System of Divinity as proper and important, as a System of Jurisprudence, Physic, or Natural Philosophy?
—Samuel Hopkins, 1793

An Essay on the Principle of Population
—Thomas Robert Malthus, 1798

ENLIGHTENMENT AND PROBABILITY—KNOWLEDGE BEFORE THE DISCIPLINES

From roughly Ephraim Chambers and his *Cyclopaedia* in the second quarter of the eighteenth century to Adam Smith and *The Wealth of Nations* at the end of the third, the knowledge we now label Enlightenment was shaped by the genre of system—particularly by efforts seen as both monumental and monstrous to scale up systems into Master Systems. In the last two decades of the century, however, that strategy gave way to the different uses of system evident in the quotations above: specialized systems scaled to old and new institutional rubrics (Divinity, Jurisprudence, Physic) and the dispersing of systems into forms other than itself, particularly the specialized essays that helped to form the modern disciplines (the principle of population in politics and economics). To best understand why this happened, how, and the consequences, we need to flesh out more fully what I have been calling the *effects* of making Master Systems.

Adam Smith's casting of that effort as a nation-building project highlighted how the "arranging and methodizing" that built a System out of systems could be understood as "useful labour." In his view, system, in comparison to other genres, was a particularly appropriate tool for that job due to two features that we have already identified: system's links to both the "simple and natural" and to the associated issues of communication and access that haunted Newton's exchange with Hooke. Smith built Master Systems so that more people could use more knowledge, a precondition, he argued, for Scotland's transformation into a commercial society. Making knowledge matter more, as with Newton's wrestling with the inclusion of the "System" book in the *Principia*, had been a central preoccupation of the Enlightenment effort to know the "World." By Smith at midcentury, that preoccupation had become a personal and national occupation: the making of a kind of knowledge that could matter more—that could *work* in that world.

The means to that end for Smith and his cohort was to use the inclusiveness of Master Systems to construct a common ground on which to win assent and actions from others. In the most remarkable passage in his remarkable book on Adam Smith, Nicholas Phillipson describes Smith's method in his Edinburgh lectures as

> designed to explore propositions that appealed to the common sense of an audience, in the hope of showing how they could develop credible maps of the world of knowledge. It was Smith's first exercise in presenting philosophy as a matter of mapping the world of experience as contemporaries knew it. How far those maps could be said to represent the geography of what was true rather than what was plausible or probable was a matter that a Humean like Smith chose not to explore. (Phillipson 2010, 100)

Yes, we are the heirs to Enlightenment, but the Enlightenment was different in ways we are only just beginning to grasp.[1] For Smith, true knowledge was useful knowledge that worked in the world to change that world. For us, knowledge is knowledge because it is true—because it bears some disciplinary-specific relationship to truth, whether conceived in absolute or relative terms.[2] Whether that knowledge is useful is another matter that varies according to each discipline's expectations and practices. Calls on both sides of the Atlantic in the early twenty-first century for measuring the impact of knowledge have thus been met with varying levels of skepticism or resistance.

For Smith, having impact was not an additional concern of knowledge making or a bonus for doing it well, but the point of doing it at all. And what enabled impact was the sharing of credible maps of knowledge with his contemporaries. Thus, before mapping his major Master Systems, such as *The Wealth of Nations*, he labored first at constructing a concept that could bear the load of credibility. "Sentiments," the theory of which was Adam Smith's first and foundational text, were not—as most often presented—feelings, even refined ones, but "probable behaviors"—thoughts and actions posited as so likely to be shared that they formed a "world of experience" for him and for others—a world that knowledge could map and in which it could work. For Smith, knowledge was itself unknowable without sentiments. Sentiments mediated the very possibility of knowledge. Knowledge could have impact *on* the world because it was *of* the world.

In the history of ideas rather than of mediation, sentiment and its cluster of associated terms (*sentimental, sensibility*) have embarrassed narratives of an "age of reason"—and have thus tended to be ignored, cordoned off from Enlightenment (as a feature of "early" Smith or a separate "Age of Sensibility"), or transposed into a binary of "reason" versus "emotion." But in Smith's "as if" Newtonianism, sentiments had the same necessary, fundamental, and innovative role in systems of moral philosophy as mathematical principles had in natural philosophy. Just as, for Galileo and Newton, the book of nature was written in the language of mathematics, so, for the moral philosophers of the Enlightenment, the book of human nature was written in sentiments or their functional equivalents.

That parallel has been difficult for us to recover for two reasons. First, the acts of translating these languages into knowledge that could be read by others required different strategies yielding different—to use Newton's term—"style[s]." Thus for Newton, worrying how to best communicate the *Principia*'s "System" of the physical world, translation into the mathematical style required, in Catherine Packham's words, "a rigid opposition between demonstrable and certain knowledge on the one hand, and wanton fancy on the other" (Packham 2007, 169) For Smith, however, worrying how best to convey Master Systems of human nature, translation into a style amenable to sentiment required not Newton's binary but—again in Peckham's formulation—"a model of projected, possible, and credible knowledge, whose acknowledged contingencies were balanced against the value of its social effects." To suggest however, as Packham does, that one version of knowledge "replaced" the other is to overlook the generic foundation they both shared—how knowledge in both cases was shaped by the making of Master Systems.

That continuity has also been difficult to recover because sentiment's role in the scaling up of system was historically specific to the Enlightenment—to the geographical, demographic, and institutional conditions of the mid-eighteenth century in the West. Making Master Systems was a plausible project for Smith and for others as long as the sample size for positing probable behaviors remained within manageable limits. Another way to put this problem is to pose it as a question to ourselves: What can perform the legitimating, common-ground functions of sentiments today? If we want knowledge to matter to us and to others, what can take their place now?

We cannot recycle Smith's "sentiments" today because they don't scale up to current conceptions of "we"—in numbers or in scope (i.e., diversity). This is not surprising, since his method of sampling for probable behaviors was local and anecdotal, even as his contemporary, Thomas Bayes, was establishing the mathematical basis for computing probability. Bayes's enterprise provides a useful counterpoint, for mathematical principles, unlike sentiments, have scaled up in strikingly successful ways. Mediated by new technologies of computation—larger memory, faster processing—probability has emerged as once again central to knowledge but in a startlingly different way. While Smith derived principles such as the division of labor from conjecture—"people will probably behave this way"—we now posit certain forms of behavior as themselves probabilistic.[3]

For Smith, the search for the probable was an adventure in system from start to finish. It began with his identification of system as one of the most probable of human behaviors. The "invisible hand" that Smith famously placed at the center of political and economic debate—mesmerizing, gratifying, and enraging generations of critics, politicians, and economists—turned out to be animated by a very visible eighteenth-century form. In the paragraph after he first played that hand in the *Moral Sentiments*, Smith tipped it in a bid to explain why it worked to ensure equal distribution of "the real happiness of human life":

> The same principle, the same love of system, the same regard to the beauty of order, of art and contrivance, frequently serves to recommend those institutions which tend to promote the public welfare. … From a certain spirit of

> system … from a certain love or art and contrivance, we sometimes seem to value the means more than the end, and to be eager to promote the happiness of our fellow-creatures, rather from a view to perfect and improve a certain beautiful and orderly system, than from any immediate sense or feeling of what they either suffer or enjoy. (Smith [1759] 1976, 184–185)

What astonished Smith in Newton—the desire and capacity to improve system—was, he asserted, a fundamentally probable behavior of mankind. It was a sentiment even more fundamental, he argued, than sympathy, the behavior on which most Smith commentary has focused. This sentiment for system works collectively and thus invisibly to produce systems in all areas of human endeavor, from small-scale interactions to the institutional.

As probable behaviors, Smith's sentiments function as principles of action that underlie the formation of systems. And the more systems compiled within a Master System, the higher the probability of a behavior carrying the epistemological weight of a sentiment. Thus the enabling premise of part 7 of his *Theory*—the final part in which he characteristically incorporated all other systems into his own—that "from some one or other of those principles which I have been endeavouring to unfold, every system of morality that ever had any reputation in the world has, perhaps, ultimately been derived" (Smith [1759] 1976, 265). The confidence that yielded "every"—qualified only mildly by the "perhaps"—was the explanatory confidence of Master Systems, the effects of which generated the sense of ambition and progress later labeled "Enlightenment."

But whether we find that confidence admirable or suspicious, we should not erase the differences of that knowledge project from our own. Discussions of Smith and his cohort, especially those conducted under the soothing continuity of apparently persistent ideas, most often ignore or underplay those differences. Andrew Skinner has helped to counter that tendency by highlighting Richard Olson's analysis from 1975:

> The great significance of Smith's doctrine is that since it measures the value of philosophical systems solely in relation to their satisfaction of the human craving for order, it sets up a human rather than an absolute or natural standard for science, and it leaves all science essentially hypothetical. Furthermore, Smith implied that unceasing change rather than permanence must be the characteristic of philosophy. (Olson 1975, 125)

"It was exactly that perspective," Skinner adds, "that led Smith to take the bold and novel step, in an age dominated by Newton, of reminding his readers that the content of that system was not necessarily true" (Skinner 2001, 15).[4]

Knowledge of that kind, as Phillipson puts it, "is better described as a form of understanding, to be considered in terms of the ideas and sentiments we acquire in the course of common life." Just as Smith understood all forms of morality as systems arising from sentiments, so Hume explained every religion as a system formed around its informing "miracle"—the distinctive principle that provides its members with shared beliefs (Hume [1748] 1975, 109–131). With that form of understanding operating through visible and invisible hands, the domain of moral philosophy, with its forms of learning and of belief—like the domain of natural philosophy with its planets, moons, and comets—had been transformed. Bacon's handshake was sealed with system.

Less than a century after Newton reluctantly published his system of the physical world, the intellectual world of the West had filled with systems.

AFTER MASTER SYSTEMS—THE REORGANIZATION OF KNOWLEDGE

For those who chose system as a vehicle for change, the effect of their efforts was rapid, obvious, and rewarding. "The famous *Encyclopédie* of the French philosophers," notes James Buchan, "had devoted a single contemptuous paragraph to Écosse in 1755, but by 1762 Voltaire was writing, with more than a touch of malice, 'today it is from Scotland that we get rules of taste in all the arts, from epic poetry to gardening'" (Buchan 2003, 2). Smith's master plan for Master Systems had clearly succeeded, but that enterprise quickly became a victim of its own success—and due to what Newton would have called a mathematical principle: more writing of more and more systems made reconciliation into a single SYSTEM less and less likely—there was simply too much system for System to master.

Efforts continued to be made, of course, but they tended to sink under their own weight. In 1798, for example, William Belcher wrote *Intellectual electricity, novum organum of vision, and grand mystic secret.* Described in the subtitle as "an experimental and practical system of the passions, metaphysics and religion, really genuine," Belcher's claim to the real was grounded in the same generic procedure that served Smith so well: he embedded extracts of other systems by, in order, "Sir Isaac Newton, Dr. Hartley, Beddoes, and others." Although not explicitly critiqued, but rather cited positively, those systems are all generically subordinated within the "grand" scope of Belcher's own "secret" (Belcher 1798, 1).

Intellectual Electricity is still a secret to most of us two hundred years later, but—we can see with hindsight—the odds were stacked against it. Not only did it face a proliferation of systems that challenged the genre's scalability; that proliferation was itself part of a rather sudden and startling increase in all forms of print in Britain. Far from rising steadily throughout the century, print production, as noted earlier, did not take off until the last two decades of the eighteenth century. The printing press had, of course, been invented much earlier, and its products had already been appearing in quantity and in force. But that fact only sets the stage for the pertinent question for historical analysis: the question of the relationship between quantity and effect—at what point was more *different*?

The magnitude of this end-of-the-century increase can help us to refine that query. When, we might ask, does proliferation turn into saturation? At the level of the pervasiveness of the technology itself, the term *saturation* can point to the moment in which new norms of practice take hold. In that sense, saturation is signaled by the paradox of access. On the one hand, *saturation* means that more people have more access to the technology; on the other, it signals that, strangely enough, direct access is not required—that even those lacking or refusing access are transformed by the ubiquitous presence of the technology. This is the tale now being retold by the early twenty-first-century advent of electronic and digital media. Whether individuals have that technology or try to avoid it, there is now a pervasive sense that there is nowhere to hide from the differences it generates; in fact, the desire to hide is itself an index to saturation and confirmation of change.

Late eighteenth-century Britain is thus a laboratory for studying the consequences of saturation for both particular forms of print technology, such as periodicals and system, and for those who made and used those forms. Both the *Monthly* and the *Critical Review*, for example, had to abandon their prime editorial imperative of taking notice of everything entering print under the pressure of too many things. That shift laid the groundwork for new kinds of periodicals, such as the *Edinburgh Review*, which specialized in longer articles on fewer texts. Such changes point to how moments of saturation are particularly well suited to pursue the issue of "blame," since, as we are currently experiencing in our encounters with computers, saturation by a new technology often produces that effect. System in particular became a magnet for blame because, in the midst of the overall proliferation, that genre had also saturated itself.

When Systems could no longer contain the growing mass of systems, the obvious consequence was that the project of making Master Systems wound down in number and in prestige. The genre of the encyclopedia provided the first evidence of this. The French *Encyclopédie* of the 1750s followed Chambers's venture of the 1720s in aspiring—with the help of the alphabet—to the status of an all-inclusive Master System, a goal visualized in a finely scaled tree diagram of the branches of knowledge. The tree stood for the unity of philosophical inquiry. Each subject-specific effort to be comprehensive extended a branch by turning previous systems into parts of a new one, enacting the same logic of scalability—of systems within systems—that shaped the tree.

When the first edition of *Britannica* appeared between 1768 and 1771, it too accumulated previous systems; however, they were not mastered in the same way. The editor even took the trouble to explicitly denounce alphabetical order and the very idea of a comprehensive diagram. Instead, as we saw earlier, systems were embedded in a hybrid form that proved to be *Britannica*'s most striking formal innovation. Materials from all of the sciences and arts were "digested" into substantial treatises that formed a now familiar "arrangement" of subjects resembling the modern curriculum; in effect, the encyclopedia became the home of protodisciplinary systems (*Britannica* 1771, v; *Britannica* 1797, x).

Per the Tectonics argument I advanced earlier, as *Britannica* grew in size, not only did the number of protodisciplines increase, but the boundaries between them thickened. On the new generic platform of system plus history, what had been only digested was now defined and deepened. Each of the "distinct" new forms of knowledge was now packaged in the third edition with its own "History, Theory, and Practice." All of these nascent disciplines, as I argued in chapter 2, instituted themselves as "DETACHED PARTS OF KNOWLEDGE" by telling their own histories.

For William Belcher, who published his "grand mystic secret" in 1798, a year after *Britannica*'s third edition, the various parts of knowledge were still very much attached to each other. But the Master System project authorized by that sense of attachment had by that time already been largely eclipsed. Under the pressure of proliferation, it gave way to new practices, not only of encyclopedic detachment but also of variations of the practice that I am calling embedding. Genres, that is, had to be combined in new ways. Systems had been and continued to be embedded—wholes becoming parts of other wholes—but the containing genres were increasingly not themselves systems. Instead of systems becoming parts of Systems they were inserted into other forms.

Although this may sound like a mere technical shift in the use of a single genre, we have already seen how the deployment of system in relationship to itself and to other genres—as in Galileo's message, Newton's treatise, Chambers's cyclopedia, and Smith's System—altered not only its own status but the nature, shape, and purpose of knowledge itself. The end result of this particular shift in the use of system was nothing less than to enact the turn into modernity presaged by *Britannica*'s editions: a transformation in the ways of knowing from the Enlightenment organization of knowledge in which every kind was a branch of philosophy, moral or natural, into our present organization: narrow but deep disciplines detached from each other and then divided between what we now know as the humanities and sciences (Siskin 1998, 79–81).

Let me recapitulate this shift in procedural terms with some dates. The generic marker of Enlightenment was the monumental effort to enact and preserve the philosophical unity of the branching model by scaling systems up into Master Systems—a project that persisted from roughly 1740 to 1780. During the last two decades of the eighteenth century, however, that procedure gave way to the transformations that shaped the modern disciplines: the dispersing of more and more systems into other forms and the scaling down of systems into vehicles for the specialization and professionalization of knowledge.[5]

SYSTEM AMONG THE DISCIPLINES—EMBEDDEDNESS AND TRAVEL

To grasp system's role in this reorganization of knowledge, we need to examine the effects of particular types of embedding. When dispersed, for example, into essays—a form characterized back then not by comprehensiveness and completion but by irregularity and fragmentation (Johnson's "loose sally")—the genre of system functioned in a new way. Instead of producing more knowledge by scaling up into more encompassing versions of itself—its authority amplified by its own inclusiveness—system had to face the problem of fit: of how well a system can carry its authority into a form different from itself. Can it travel?

Traveling of this type was an act of discipline in that the destination in each case was what we now call a discipline. Knowledge thus grew through efforts increasingly concentrated within increasingly stable precincts of inquiry. The primary purpose of that knowledge was no longer to work in the world but to work within those boundaries, probability (fitting into the world) giving way to a will to truth (fitting into the discipline).[6] Disciplines in this newly modern sense thus became narrow-but-deep arenas for specialization. The work of knowledge changed, and so did knowledge workers. As philosophical branching gave way to modern disciplinarity, the established liberal professions were reformed into a new regime of expertise—modern professionalism.[7]

We are the evidence for the historical importance of that change, for we remain part of that regime and part of that organization of knowledge. But the impact of this redeployment of system was dramatically evident even back then. The text by Thomas Malthus whose title heads this chapter, and those by William Wordsworth and Mary Hays that I turn to later, all share with *Intellectual Electricity* the same date, 1798, and the same feature: they contain systems. Unlike Belcher's "grand" effort, however, they are not themselves Systems; they are, in their own words, an "essay" a "preface," and an "appeal." My focus here is on how this practice of embedding in other genres reshaped knowledge; in the next chapter, I turn—as we did earlier with Master Systems—to the large-scale effects of this change in form.

Malthus's title articulates that practice: *An Essay on the Principle of Population*. Embedded in the text is the principle of the system that figures man's fate as a "lottery" (204): "Population, when unchecked, increases in a geometrical ratio. Subsistence increases only in an arithmetical ratio" (18). This is the phenomenon of population reduced to the regular ordering of system, and if it had become part—through critique or affirmation—of an encompassing Master System, then the whole work would have either followed Belcher over the top or secured some of the same kind of ongoing power as *The Wealth of Nations*. That text speaks—*whether we agree with Smith's ideas or not*—with an authority derived from a seemingly seamless link between the embedded principle of division of labor and the encompassing concern with wealth and nation.

To what is the principle of population linked? Since this essay was part of Malthus's debate with William Godwin over "the future improvement of society,"[8] its embedded principle should have visited that concern. How, for example, should we behave in the face of the famine it predicts? But the link to these moral issues was not made seamlessly in Malthus; in fact, the principle was not brought to bear fully on morality until a second edition. In a history of ideas, the initial difficulty and revision might be construed as a subject having second thoughts: did Malthus change his mind? Within a history of mediation, however, we can bring more than local possibilities into play. Was this a pattern tied not to an individual author or text but to a widespread change in generic procedure—in how the genre of system was used? Systems embedded within Master Systems, per Smith, became parts of that newly comprehensive whole. Systems embedded within essays, however, may not have exhibited the seamless self-similarity of scaled-up systems. Not only are the two genres different in kind, but essays came to be understood formally by Malthus's time as not necessarily aspiring to the wholeness of systems—as deliberately circumscribed attempts.

That was the key difference between the interrelations of system and essay at midcentury versus its close. Instead of "essay towards system"—the standard formulation we noted earlier in Burton's and Granger's efforts of the 1750s to 1770s—we see the phenomenon I am calling embedding: "systems in essays." The ambition of "towards"—which fueled the (over)production of Master Systems—gave way to a very different gesture: Malthus's text begins not with aspiration but with an apology. His work was an "essay," he emphasized, and "might, undoubtedly, have been rendered much more complete" (ii). Completion, that is, was certainly an option for Malthus—and one he clearly needed to acknowledge—but it turned out not to be the only option. When he revised this essay, it stayed an essay. This meant that embeddedness came to work in a new way. As we have seen, systems and features of systems had mixed with and been embedded in other genres in the past. But the effects of doing so changed when embedding systems was no longer a prologue to completing them but an end in itself.

If we think of incompleteness as a generic feature, we can now use it as an index to the changes in the shape and organization of knowledge I am trying to describe: incompleteness worked in different ways at different historical moments. We have seen how it played a key strategic role in Bacon, underpinning his argument that fragmentary forms such as essays and aphorisms were necessary starting points for new knowledge. In the enterprise of "essay towards system," incompleteness signaled not a starting point but a goal; it established for Burton and Granger the formal intention to construct Master Systems. Apologies like Malthus's, however, allowed for a turn from that intention; incompletion could occasion a nod in the direction of Master Systems but now only as a polite prelude to taking on a different task.

In retrospect, the reasons for this last shift were twofold. First, in that late-century moment of print's proliferation, delay in joining what Malthus called the "conversation" in print risked losing one's place in it. In his case, having already suffered a "long and almost total interruption" in publishing his work, he opted for incompletion rather than further delay (ii). And, second and most crucially, incompletion was not the liability it had been when Master Systems were the primary mode of completion, turning all other conversations into threads of its own. When alternative types of "Full EXPLANATIONS" took hold—to use one of *Britannica*'s phrases for its protodisciplines—scalability as the mode of completion (wholes becoming parts of larger wholes) gave way to accumulation (adding parts to parts). Rather than having to scale up "towards" systems, essays could contribute to knowledge as discrete additions to subject-specific conversations—eventually without apology. Within the boundedness of disciplines, the incompleteness of essay became its professional virtue, conveying, for modernity, originality (adding to a field) and collegiality (contributing jointly to it).

MORE THINGS

> Not that system is by any means to be thrown aside; without system the field of Nature would be a pathless wilderness; but system should be subservient to, not the main object of, pursuit.
> —Gilbert White, 1778 (White 1789, 232)

Gilbert White offered this advice as the century entered its final two decades—the moment when Master Systems gave way to the new deployments of both system and essay that I have been describing. One hundred years after Newton helped to negotiate a new philosophical prominence for system, that genre was once again being repositioned. White's use of the term *subservient*, however, can be misleading if it suggests a falloff in numbers or ambition. In fact, the percentage of texts with variants of *system* on their title pages, as I noted in chapter 1 (figures 1.4 and 1.5), took an exponential turn to new heights in the 1790s.

Although most of those systems did not aspire to the comprehensive sense of a Master System as the model for, or container of, all other systems, the turn from that form of optimism did not mean that they were no longer ambitious. Here is where Condillac's formulation I cited earlier can be appropriated to demarcate historical change. In his mathematical metaphor, systems exhibited the feature of simplicity by maximizing the ratio of things to be explained to the principle explaining them. That ambition could be fueled by either aggrandizing the explanatory principle—Newton's "law" or Smith's System—or emphasizing the number and variety of things. In the embedded and specialized systems of the 1790s and after, the latter became increasingly common. By the changes in the mode of completion I have just described, accumulation gained ground over scalability.

Malthus and, as we shall see, Wordsworth and Hays in the very same year (1798) all followed that pattern—that is, their embedded systems did—and thus claimed for system a new kind of explanatory power. The principle of population, according to Malthus, had evaded even "the most penetrating mind" because of what had been left out of systematic consideration: "The histories of mankind that we possess, are histories only of the higher classes" (32). With the lower classes included, the regular oscillations of population could be detected. Translated from history into system, their regularity became a "law of nature" (15), turning time itself into a part of the population principle and making the future—the centerpiece of the debate with Godwin—known.

This same kind of inclusiveness was what made William Wordsworth infamous—a target for fellow poets and critics—and then famous. Embedded within *Lyrical Ballads and Other Poems* are the basic components of system's ratio: explanatory principles and "things" they explained. All of the reviewers of the 1798 edition read the poems through the frame of the Advertisement,[9] and in critiques of subsequent editions and other volumes, Wordsworth was explicitly portrayed as writing "upon" system.[10] What concerned readers most, however, was not the genre but the things it now included. The Advertisement opens by switching abruptly from the "subject" of poetry to its proper "language." The only link between them is a new sense of possibility—that "every" subject and even the language of the "lower and middle classes" are potentially poetical (Wordsworth [1798] 1965, 7). Many early critics denied one aspect or another of that potential, but kept the possibility of writing upon system open. "The 'experiment,' we think, has failed," wrote Robert Southey, "not because the language of conversation is little adapted to 'the purposes of poetic pleasure,' but because it has been tried upon uninteresting subjects."[11]

By adding to poetry the subjects and the languages of the lower and middle classes, Wordsworth had heightened the ratio of system as Malthus had—by increasing the numerator, the number of things. But that also lowered the denominator, for the discarding of traditional standards of decorum and of poetical diction left "pleasure" as the only principle underlying his "experiment." Embedded within an Advertisement—like the modern essay, a purposefully incomplete genre—the principle of pleasure faced, like the principle of population, the problem of fit and travel—of how well and far it could carry its authority. That problem was perhaps even more pressing here than in Malthus, since the first stop was the poems: did Wordsworth's system for poetry, his readers asked, productively inform his practice?

The answer for most readers was, at best, "maybe." For Wordsworth, as for Malthus, it was revision. The Advertisement lengthened into a preface—also, like the essay and advertisement, strategically incomplete. Its open-endedness only raised the stakes, challenging the system to travel even further—to explain "the revolutions not of literature alone but likewise of society itself" (243). It induced, that is, more revision, leading out of the preface itself into another form of incompletion: an appendix.

The concluding sentence of that appendix uncannily enacts the phenomenon of embedding it describes: Wordsworth wrote that he was tempted

> here to add a sentiment which ought to be the pervading spirit of a system, detached parts of which have been imperfectly explained in the Preface—namely, that in proportion as ideas and feelings are valuable, whether the composition be in prose or in verse, they require and exact one and the same language. (318)

As with the content of Malthus's morality, the idea after the dash—that poetry and prose should share the same language—garnered much attention within the then newly forming disciplines. Within a few years (1813), we find the first reference to a new kind of home for that attention: the "English Department" (Michael 1987, 377). What Wordsworth argued before the dash, however, has received less attention: his insistence that system "ought" to reshape other genres and knowledge itself. Through the embedding of system and its "parts" into other forms, that genre remained pervasive and instrumental in reshaping what was known and how.

In the short term, Wordsworth's reputation was at stake. What emerged in the long term, however, was an ongoing field of inquiry: an extended discussion of system and pleasure, of the possibilities and consequences of mixing the two. Is being systematic conducive to pleasure? How systematic can and should the measurement of pleasure be? The conversation deepened, that is, into the modern discourse of aesthetics—aesthetics transformed, as Samuel Taylor Coleridge noted with some regret, from the original eighteenth-century sense of a "science of the senses" to matters of "taste" (Coleridge 1995, LXXV, 938).

In that conversation, the focus shifted from where it had been in Newton and Smith—on system itself and the comprehensive knowledge of Nature that system was supposed to contain—to the traveling of system into disciplines and their more local knowledges and practices. Wordsworth became the authorial locus for an increasingly professional discussion of poetry and aesthetics. Whereas Master Systems simplified by including everything in one thing, embedded systems gave us the divide-and-conquer strategy of modernity: simplification through specialization. Pleasure measuring, for example, helped to turn "literature" from the *Encyclopaedia Britannica*'s inclusive, Baconian sense of "learning or skill in letters" into the specialization we call English Literature.

Both Wordsworth and Malthus, then, paired new things with principles that had to travel to explain them. By admitting in the appendix that his own system was "imperfectly explained," Wordsworth acknowledged both the ambition and difficulties of that travel; it generated out of system a "spirit"—something powerful, pervasive, and yet amorphous. To appreciate fully how that spirit has haunted modernity, we need to enter the history of blame in Part III. Chapter 5 focuses on system and the political, while chapters 6 and 7 track system into the newly emerging categories of "culture" and the "social," including a case study of how the very "spirit" Wordsworth invoked shaped his career and reputation.

But to set—and populate—the scene for those hauntings, we need to turn first to one more example of the "more things" strategy—and its consequences. In addition to Malthus's *Essay* and Wordsworth's Advertisement, 1798 also saw the publication of an *Appeal* by Mary Hays. Addressed *To the Men of Great Britain in Behalf of Women*, it too featured an Advertisement. Hays told her readers that she had started this work "some years ago," when the subject had "some degree of novelty to recommend it," but was interrupted twice by the appearance of other works on that same subject. As with Malthus, interruption points to incompletion; not only was the project delayed, but Hays repeatedly refers to the final result as an "attempt" and a "sketch" (Hays 1798, unnumbered Advertisement 1, 2, 4).

By specifying the nature of the interruptions, she also provided a time line that makes publication in 1798 make personal—and to us, political and historical—sense. The first delay occurred exactly a decade earlier with the publication of Alexander Jardine's *Letters from Barbary, France, Spain, Portugal, &c.* Worried that her work would have "little claim to notice" in the shadow of his comments on women, she did not resume her pursuit for three years. But "just at that period"—1791—another piece appeared with a "pointed title, and declared purpose" just like her own: Mary Wollstonecraft's *Rights of Woman*.

Hays experienced twice within a few years, then, what Malthus had feared in the face of print's proliferation: losing her place in the "conversation." A "greater interval than the last elapsed," Hays tells us, for, with all hope of "novelty" lost, resumption required a new rationale. The argument that she finally constructed was generic in that it insisted on matching kinds of books to kinds of people. Her *Appeal* would not duplicate the *Vindication*, she reasoned, since different readers needed to be addressed in different ways "before the public opinion is influenced to any degree." Generalizing from the competition "to which I have alluded," Hays argued that what "raises" some works "in the eyes of the few, either sets them beyond the reach of the multitude; or, what is infinitely worse, renders them obnoxious to its hatred and persecution." The vicious reception accorded the *Vindication* during the 1790s thus provided Hays with a historically specific purpose for her work as that decade ended: 1798 becomes the moment to find new ways to "manag[e]" the "prejudices of mankind, to make new and unexpected truths palatable to common minds" (Advertisement 5, 6).

This strategy of inclusion—more things, including more kinds of people—sought to improve on the past, whether by finding a better population principle, writing proper poetry, or convincing more readers. And as with Malthus and Wordsworth, Hays gestured toward the "common" by embedding systems as parts within adventurously incomplete wholes. In "that slight, hasty, and unfinished manner, of which in some degree every part of this little work must necessarily partake" (93), Hays did not dwell on the "education of females" (Wollstonecraft's focus) but attempted to be newly comprehensive. Not only did she "write for all classes" (239), but she also addressed systematic behaviors by men (47) as well as by women (90).

The system of the sexes that Hays embedded in her *Appeal* was thus distinguished by this breadth from all earlier versions: even men, despite their privileged position, are in it. The more men try to make women into the things they want, the more women become the things they are: something that can only be described "by negatives" and certainly "not what men would have them to be" (68). Her essayistic attempt to extend the authority of systematic analysis into other things thus met the same unsure fate of Malthus's and Wordsworth's efforts. Tentatively venturing into "what women ought to be," Hays toyed with revising the whole structure of her work:

> Though I have not certainly, the vanity to believe myself equal to the talk of fulfilling the title of this chapter, to the entire satisfaction of my readers; nor what is perhaps in the first instance still more discouraging, even to that of my own; yet having once adopted I shall retain it, as it expresses exactly what I wish to accomplish, however I may fail in the execution. (125)

The result, she admitted, using generic terms, may not be a "regular system," but it will at least be a "bold … outline." Like Malthus, then, Hays tried to discipline her system into traveling further by adding more things; he revised to add another "constraint" to what men ought to do while she turned the "ought" on women. He extended his system into a second edition; she stretched her system into an outline.

RAISING SYSTEMS

> Yet, be not discouraged; exercise your understanding, think freely, investigate every opinion, disdain the rust of antiquity, raise systems, invent hypotheses, and, by the absurdities they involve, seize on the clue of truth.
> —Mary Hays, *Memoirs of Emma Courtney* (1796)[12]

Emma Courtney's advice to her nephew was not against the grain in the mid-1790s in Britain. Systems were being "raise[d]" at an unprecedented rate, including the one Hays struggled with the entire decade. Through Emma, Hays foregrounded the historical continuity in this turn to system, warning her nephew that

> a long train of patient and laborious experiments must precede our deductions and conclusions. The science of mind is not less demonstrative, and far more important, than the science of Newton; but we must proceed on similar principles.

The continuity Hays perceived lay not only in Newton's use of system but in how he used it: as a preferred form of what I have called guesswork. It was preferred because it enabled error correction; in Hays's version, raising systems also raised and exposed "absurdities," the disposal of which was a necessary but laborious and repetitive process on the road to "truth."

By specifying, however, that truths of different kinds were now at stake, this account also accommodated discontinuity: new and different "sciences" competing for "importance" might raise systems in different ways. Hays's did precisely that in print in her *Appeal* a few years later. As I have just detailed, she stretched a "regular" system into an outline to accommodate more things into a more truthful account of what men and women ought to be.

Embedding and stretching are thus two ways to describe what happened to "regular" systems as they were "raised" in unprecedented numbers as the eighteenth century drew to a close. They were embedded in other forms and stretched to accommodate more things in order to serve newly specialized disciplinary, professional, and institutional ends. System reached the statistical heights documented earlier when produced within these proliferating and expanding rubrics, especially the subject-specific disciplines associated with education and the growth of schools and in the specialized areas of expertise linked to the expansion of governmental bureaucracies, such as Pitt's income tax of 1798. The stage was set by these adventures for system to come into its own as an object of blame—to become as blameworthy as it was ubiquitous.

To capture a snapshot of that ubiquity at the turn into the nineteenth century, I engaged my research assistant at the University of Glasgow, Tom Mole, in the task of sampling systems from the library there. He searched for long-term patterns during the library's tenure as a legal deposit (copyright) institution from 1709 to 1836 and then focused in on particular examples from the latter part of that period. He identified many instances of the embedding and generic mixing I have described, noting what we might call system's viral relationship to dialogues, sermons, and, particularly, letters. In certain instances, the embedding entailed so many layers and add-ons that a more dramatic term—perhaps *swallowed*—may be in order.

Note the fate of system, for example, in John Payne's 1791 extravaganza entitled

> *Universal Geography formed into a new and entire System; describing Asia, Africa, Europe, and America; with their subdivisions of Empires, Kingdoms, States, and Republics: The Extent, Boundaries, and remarkable Appearances of each Country; Cities, Towns and Curiosities of Nature and Art. Also giving A general Account of the Fossil*

> *and Vegetable Productions of the Earth, and of every Species of Animal: The History of Man, in all climates, regions, and conditions; Customs, Manners, Laws, Governments, and Religions: the State of Arts, Sciences, Commerce, Manufactures, and Knowledge. Sketches of the Ancient and Modern History of each Nation and People. To which is added, A Short View of Astronomy, as connected with geography; of the Planetary System to which the Earth belongs; and of the Universe in General* (London: printed for the author, 1791).

Here the adding of new generic layers—an "account," "sketches," a "short view"—not only propels system's travels in protodisciplinary directions such as geography, the history of man (anthropology), and astronomy. It also enacts the strategy of adding "more things" to system while simultaneously exercising system's scalability in ever more dramatic fashion, culminating at neo-Newtonian heights: "the Universe in General."

This practice of appending other things to systems was sufficiently widespread, Mole found, to be frowned on in some quarters, where it seemed to some to be simply padding. When, for example, Benjamin Booth published *A Complete System of Book Keeping, by an Improved Mode of Double Entry* (1789), he pointedly asserted that:

> It would have been an easy matter to have introduced a variety of other subjects into this book, so as to have doubled its price to the purchaser: but, at the best, I entertain no very high opinion of the trade of book-making; and I should deserve the severest censure, if I were to pursue a conduct, which I have always been so ready to condemn in others. (6)

To track these permutations of system is to recognize that the proliferation of print in the late eighteenth and early nineteenth centuries was multidimensional; it was not simply a matter of more texts but also of proliferations in size, kinds, topics, and price range.

The scalability of system thus contributed to the scaling up of print and its roles in society. Raising systems became increasingly a matter of practical applications, with the genre serving a perceived need for increased efficiency, accountability, and regulation in the administration of increasingly professionalized work both at home and amid growing global networks of trade. Crucially, for system makers, needs of this type were not static. To sustain increases in efficiency, for example, the systems themselves had to change, and that was where system's capacity for error correction came into play. Not only could new systems replace old ones, but each system could—in fact, needed to—supersede itself through new editions.

That was the fate of William Mavor's 1785 work:

> *Universal stenography; or a new complete system of short writing: Rendered perfectly easy to read & write; freed from all prolixity and obscurity; adapted to every purpose in which short writing is useful or ornamental and attainable in a few hours by the most common capacity: Being an improvement on the most celebrated systems that have been exhibited to the public for above a century past and superior to all in ease, elegance and expedition. Designed for the use of schools & private gentlemen.*

"Designed" to be a professionally useful culmination of a century's worth of system making, Mavor's effort improved itself through ten editions over thirty years starting in the 1780s.[13] It also launched improvements by others, systems begetting supplementary efforts to further systematize themselves, as in James Duncan's *A New Introduction to Stenography, or Short-Hand Writing; Being an Attempt to Facilitate the Acquisition of Dr. Mavor's Standard System, by a more simple elucidation of its elements* (1816).

In similar fashion, Alexander Ingram's *Concise System of Mathematics in Theory and Practice for the use of Schools, Private Students and Practical Men* reached its eighth edition in 1816, thanks in part to the appearance, from the same publisher, of James Trotter's *A Key to Ingram's Concise System of Mathematics; Containing Solutions of All the Questions Prescribed in That Work*—a work that itself went through four editions by 1843. In similar fashion, when George Jackson published *Jackson's New and Improved System of Mnemonics, or Two Hours Study in the Art of Memory* (1816), he included on the recto facing the title page an advertisement for a "Chart of the World [...] upon the system taught in this work," available from his bookseller.

Musicology, Mole found, became a particularly fertile field for this form of proliferation—printed systems generating ancillary forms of print and more systems. In the Preface to his *New Theory of Musical Harmony, According to a Complete and Natural System of That Science* (1806), Augustus Frederic Christopher Kollmann detailed the generic trail that led to his new work:

> Ten years ago I laid before the Public my first Theoretical work, entitled *An Essay on Musical Harmony*, which was founded on Kirnberger's system, as the best then known. But having since discovered the cases in which that system

> is still incomplete, and imperfect, I have endeavoured to remove them; and the result has been the present Work.
>
> This I first endeavoured to publish as a mere addition to the said Essay, and announced it three years ago, under the title of a *Key to a complete and natural System of Musical Harmony*. But in arranging the materials collected for it, I found that I could not unite the two Works, without spoiling them both; and that it would be proper to show the application of the new system to every branch of the science, in a work by itself, and independent of the former one, as I now presume to offer it in the following sheets. (i)

Unlike the claims to completeness made by the Master Systems I described earlier, Kollmann claimed "Complete[ness]" in a manner consonant with specialization. As narrowing displaced inclusiveness, these contracted systems made room for more of their own kind. Readers of Kollmann wishing to understand any other areas of music than harmony had to turn elsewhere. Taking advantage of these systematic delimitations, Kollmann directed those interested in "the nature and form of musical pieces" to one of his earlier efforts, an *Essay on Practical Composition* (1799). Thus a "system" (by Kirnberger) produced an "essay," which led to a "key" that stretched into a "system," which pointed back to another "essay."

The raising of more systems—and their interrelated genres—was but one effect of specialization. Narrowing but with strong boundaries also transformed systems into a form of property that could be made legally off-limits to others. Edward Thomas Jones not only published *Jones's English System of Book-keeping* (Bristol, 1796); he also patented it. To devalue that property, a competitor named James Mill targeted "Mr. Jones's new or ancient, or bifold or trifold, or simplified or amplified System" (6) in a pamphlet called *An Examination of Jones's English System of Book-Keeping* (1796). The need to defend the value of his property led Jones, in turn, to publish a rebuttal in his *A Defense of the English System of Book-Keeping* (1797).[14]

Increasingly complex elaborations of system's new relationship to property quickly followed. Thomas Papps not only took out a patent on his system, but developed licensing strategies to protect the income he derived from selling the books that explicated it and enabled its use. Papps began the second volume of *An Improved System of Book-Keeping, Being the*

Introduction to Papp's Patent Improved Account Books (1818) by complaining that his system had been pirated. This was not a reference to his book being printed without permission, but to his system being employed without license *and* without purchasing the special account books that he had manufactured for it. As a rearguard action, Papps supplied a list of prices for "a license for adopting the System without purchasing the Patent Account Books, and also for the privilege to Manufacture them" (unnumbered page). Other strategies for extracting a profit from the specialization of systems took a pedagogical turn, such as establishing official classes. Charles Morrison's *Complete System of Practical Book-Keeping, Applicable to All Kinds of Business* (1822) includes an advertisement for his services as a teacher of his own bookkeeping system.

The entire endeavor of education itself became one of the most important sites for the raising of systems. Those included not only more specialized systems of education but also more systems for the teaching of other topics. In his *System of Geography, for the use of schools and private students, on a new and easy plan* (1816), Thomas Ewing presented systematic geography as a pedagogical tool for avoiding embarrassment in a new age of news from abroad:

> To read of places and events, of which we know neither the date nor the situation, is nearly as mortifying as to meet with words of which we know not the signification. Even a common newspaper must be very imperfectly understood by him who is unacquainted with Geography. (5)

More information not only meant more systems but also a growing need to make them more accessible. Worrying that his subject was "remote from general apprehension" (xi), Francis Ludlow Holt added "a more brief and intelligible synopsis" (xii) to his *System of the Shipping and Navigation Laws of Great Britain* (1820).

The raising of systems at the turn into the nineteenth century was thus a two-pronged phenomenon. Raising was a matter both of number—more and more specialized systems—and of value—to their makers and to navigating everyday life. Glasgow Library yielded considerable evidence for both. It inspired Mole to switch genres from his bibliographic note taking to narrative—an imaginary biography of a new kind of citizen, Systematic Man:

Having been educated at infant school, where his tutors might have learned from *A manual of the system of instruction pursued at the infant school, Meadow Street, Bristol* (by D.G. Goyder, 1825), and at one of the new Sunday Schools, where he might have studied *The complete duty of man: or, a system of doctrinal and practical Christianity* (by Henry Venn, 1759), he would have progressed to a systematic education at grammar school. His school might have followed an old system, such as *The Eton system of education* (1834) or a newly imported one such as that contained in *A Manual of the system of discipline and instruction for the schools of the Public School Society of New-York, instituted in the year 1805* (1845).

If he'd gone to university, he would certainly have encountered weighty intellectual systems such as Newton's, but suppose he got a job in a commercial business instead. He could have made up for any failings in his education by reading *A complete system of mental arithmetic, embracing every variety of commercial calculation necessary for the counting-house and shop,* by James Ferguson (1835), or by learning *An entire new system of mercantile calculation* (Anon. 1795). He might have learnt his trade through Benjamin Booth's *Complete System of Book-Keeping, by an Improved Mode of Double Entry* (1789), or *The English System of Balancing Books: Being an Improved Method, Applicable to Every Mode of Book-Keeping by Double or Single Entry* by Edward Thomas Jones (1821). He might also have benefited from *Taylor's System of Stenography* (1832) as well as the latest edition of Mavor's *Universal Stenography* (1820).

If it all went wrong, he might have needed John Paul's *A System of the Laws Relative to Bankruptcy* (1776). Imagine that he chose to seek his fortune abroad—he might have consulted *A Brief Outline of the Existing System for the Government of India* by Alexander Annand (1832), *A Compendious System of Bengal Revenue Accounts* by Francis Gladwin (1790), or *A System for Planting and Managing a Sugar Estate* by Thomas Roughley (1823). Meanwhile, the home front would have been kept in order with the aid of *A New System of Domestic Cookery; Formed Upon Principles of Economy: And Adapted to the Use of Private Families* (Anon. 1824), or Martha Taylor, *The Housekeeper's Guide; or, a System of Modern Cookery* (1831). And he might have maintained a well-regulated and pious family, reading to them from Richard Warner's *Sunday-Evening Discourses: Or, a Compendious System of Scriptural Divinity, for the Use of Households* (1828).

This tale includes but a small sampling of the ways in which specialized systems saturated the social. That penetration was paced by system's spread through the process of embedding I have been describing. Embedding more and more systems into other forms had the effect of embedding system itself into everyday life. In a striking example of this generic strategy at work in the world, David Stow wrapped system in another genre to map the flow of knowledge across multiple generations. In *Infant Training: A Dialogue, explanatory of the System Adopted in the Model Infant School* (1833), Stow turns a grandmother's visit to her grandchildren's school into a tale of two genres. She inquires about their welfare by inquiring into the system of which they have become parts. And she accesses that system by entering into a dialogue with the schoolmaster. By the first half of the nineteenth century, raising systems and raising (grand)children went hand-in-hand.

Part III CONNECTIVITIES

SYSTEM AND THE INSTITUTING OF MODERNITY

5 BLAMING THE SYSTEM—INSTITUTING THE POLITICAL

> The publick must take notice, that the leaders of that party, who have been for these last ten years involving us in confusion by displaying all the defects of our parliamentary system, and labouring to bring that system to the ground, aware as they are that that system is the cause of our distraction, now take the other side of the question, and in supporting the system are labouring to perpetuate that confusion, which by attacking and exposing the system they created, *valeat quantum* [for what it's worth].
>
> —Thomas Grady, 1799[1]

SYSTEMATIZING—SYSTEM AT WORK IN THE WORLD

The subject of the parent-teacher conference that ended chapter 4 was system. It, rather than the students or their master, was the primary candidate for praise or blame. Like the "world" to Newton, "education" was for this grandmother a system. As that mode of engaging things spread, systems mediated human behavior in more and more specific ways. Stow's embedding of system in dialogue enabled him to prescribe not only how education should work—for students and for the master—but also how a responsible grandmother should behave. In similar fashion, Malthus's embedding of system in essay, as we are about to see, led him to do the same for every "man."

System was the genre that mediated the relationship between the two modern meanings of discipline: the shaping of knowledge into narrow-but-deep areas of expertise and the disciplining of the subjects constituted by doing so. All of the newly "detached parts of knowledge," to use *Britannica*'s terms for its protodisciplines, drew their enabling boundaries by using system to claim their shares of expertise in the human.

The word we use to describe this kind of shaping and reordering—systematize—first appeared in English in the late eighteenth century. The earliest *OED* examples described how "general notions" were "gradually systematized" (1767) and how "rules" were "systematized in form" (1780). Today, we see no need to add the words "in form"; describing rules as systematized would be taken as sufficient. Untethered to any history of system as a form, we do not hesitate to deploy *systematize*—as in the example of "memory systems" offered earlier—without regard for time or place.

The benefit of taking both of those contexts into account is to better understand what was at stake in coining the term in the first place. What generated the need for a specific word to describe the act of putting other things into the form of a system was more systems—the proliferation of that form. And not only were there more systems in the late eighteenth and early nineteenth centuries; there were more of them within and attached to other forms. Those embedded and hybrid systems, I have been arguing, were increasingly more mobile; instead of scaling up into Master Systems, they traveled in different disciplinary directions, gaining explanatory power by including more things.

The word *systematizing*, then, entered the language as shorthand—or "short writing," as they would have been called back then—for all of the historically specific ways that more of everything was put "in form." Those "ways" included the modes of proliferation just described, and the "more" included entities of all kinds: human activities and behaviors, as we have just seen, as well as objects. Putting them "in form" was—per the *OED*'s first use (1811) of the word *systemization*—a reordering: "placing the several denominations," in Jeremy Bentham's phrase, "in systematic order."

That adjective, *systematic*, did appear earlier in the eighteenth century, though infrequently,[2] for it marked things as displaying characteristics of system. What these variations of *system* at the century's end—*systematize*, *systematizing*, *systemization*—signal is not just the presence or increasing number of things called "system," but the spread of a social practice: of using system's scalability to provide order within and across more things. As those things were put into the form of system—their parts reordered into new wholes—those wholes became parts of a larger whole. The reordering, that is, replicated itself at a larger and larger scale, producing a historically specific effect: that extraordinary sense of the social—of the totalizing of the social—that we now call "The System."

Part III of this book, then, discovers the social incarnations of system—the ways that system re-formed society itself, ranging from politics to clubs to cultures to the subjects that constitute them. It situates system as a form that mediated modernity, and thus it highlights connectivities as well as differences between past and present. And it explores, in this chapter, the cumulative effect of those incarnations, explaining how we came to blame The System.

WHEN SYSTEM BECAME AN OBJECT TO BLAME

In our histories of system, this is the third primary intersection with the history of blame:

- The first encounter is best envisioned in spatial terms. Blame accompanied system because system was something made—made to approximate the real and thus always expected to fall short of it. In that sense, blame had marked a gap in knowledge as the space in which Enlightenment *could* take place—it sets the scene of Enlightenment as an effort to remediate that gap and know the world.
- As an important tool in that effort, Newton and others used system to negotiate the relationship between complexity and simplicity. Since that relationship required ongoing adjustment, system often took the fall as the progenitor of either the overly simple or the monstrously complex.

At both of these intersections, system was targeted for its methodological failings—falling short or simplifying too much or too little. But as system's social incarnations—its embodiments of specialized expertise and discipline—increased in number and visibility, system was targeted not just for its ambitions but as an object.

At one extreme of this objectification was the almost metaphysical sense of The System I have just noted—the sense of being haunted by, and even a part of, something that is both ubiquitous and blameworthy. But system was also incarnated in a growing body of enterprises that took substantial physical form. Systemization in the late eighteenth century went hand-in-hand with institutionalization. In the earlier senses of "institution" as both the action and result of "instituting or establishing" (*OED*, 1a) the two terms had occupied the same semantic ground. Both referred to "the established order by which anything is regulated"; in fact, the *OED* (2b) offers *system* as a synonym for *institution* in that sense.

When "institution" took on its more modern, specialized meaning at the turn into the nineteenth century—

> an establishment, organization, or association, instituted for the promotion of some object, esp. one of public or general utility (*OED*, 7a)

—*system* scaled up from being a synonym to supplying principles and purpose. The 1790s and the opening decades of the next century saw an explosion of this form, from the Royal Institution of Great Britain, the British institution, and the London Institution, to the Plymouth Institution, the Liverpool Institution, and the Royal National Life-boat institution. Institutions of this new type were often cast as parts of encompassing systems that acted as contexts for their particular tasks. Thus a parliamentary debate in 1817 praised "our system of public schools and universities" for the manner in which its "institutions" provided "a due supply of men, fitted to serve their country" (Canning 1817, 30).

Objects of praise could, of course, easily become objects to blame, particularly when they are understood to have the power to make men "fit" their purposes. Returning to Malthus can provide us with a dramatic example of how system turned into The System. We saw earlier that the embedded systemic principle in his *Essay* posed questions of behavioral fit. How, for example, should men act in the face of the inevitably recurring famine that principle predicts? But the link to these moral issues, I emphasized, was not easily made, requiring a second edition.

The difficulty was, in terms of the reshaping of knowledge I have described, a matter of how embedded systems adjusted to their new disciplinary homes. To join these new kinds of specialized conversations, systems had to do more than—per Kevin Kelly's observation cited earlier—talk to themselves. "A thermostat system," he argued, "has endless internal bickering" about whether to turn the furnace on or off (Kelly 1994, 125). Could the commands of embedded systems (on or off) carry their authority into the disciplinary adventures of essay? Could the principle, to employ the metaphor I introduced earlier, *travel* to produce more knowledge and fit men to it?

And once on that journey, could it manage as well the change in terrain signaled by Malthus's insistence on "truth" (1798, ii) as the basis for that knowledge? That insistence was a disruptive departure from knowledge based on probable sentiments per Smith. The severity of that disruption was underlined by Godwin's claim that Malthus's *Essay* "completely reversed" everyone else's shared assumptions about the "state of the human species" (Godwin 1820, 616). Could system, that is, move from truth to truth without the shared ground of probability? If it did, how did that change the *experience* of system—the sense of its efficacy, numbers, location, and the influence it wielded?

In Malthus, this traveling question posed the particular problem of whether his particular thermostat—the on and off of growth governed by the population principle—could and should extend into and re-form human behavior—specifically, the moral issue of whether men who find themselves naturally turned on—"the passion between the sexes is necessary and will remain nearly in its present state" (11)—should, with the misery of overpopulation in mind, turn themselves off. The answer was that the commands traveled, but not all that well. Malthus had to revise the *Essay*'s morality substantially between 1798 and the next edition in 1803. Fitting system into essay in order to make men fit required ongoing work, as reflected in the change in the subtitle: *As it Affects the Future Improvement of Society* became *or a View of its Past and Present Effects on Human Happiness*. In this edition, following the vector of accumulating more things, Malthus turned from describing affect to enumerating specific "effects." To "misery and vice" as effects of the needed checks on population, Malthus added "moral restraint" (Malthus [1798/1803] 1992, 23).[3] This revision may at first appear to be only a change in content—Malthus adding a new idea to his argument—but it was the change in form I have just described that established the need for revision in the first place.

That need generated not only specific revisions but also a specific effect. Just as the sense of explanatory power attributed to *The Wealth of Nations* arose, as we saw earlier, from the process of embedding systems within Master Systems, so the embedding of systems in essays produced its own formal effect: the certainty of system extended into essay resulted in a sense of expansive but attenuated authority. In terms of the travel metaphor, this was the sense of an adventurous but difficult journey never quite at an end. System now worked both too well—the answer venturing into all kinds of questions—and not well enough—it didn't quite fit.

This was yet another variation of the falling-short scenario. When something repeatedly raises expectations but then repeatedly fails to quite fulfill them, it becomes an object of blame, habitual blame. And that is precisely what became the identifying marker for the very special effect that I am arguing emerged from the proliferating use of embedded systems at the end of the eighteenth century. That effect became a central experience of modernity: the experience of "The System" as something works both too well—"you can't beat The System"—and not well enough—it always seems to "break down." Its power is indexed by our capacity to find it everywhere and blame it for everything. The more systems we use, the more we're convinced that the System uses us: we can't beat it because we feel we're part of it, leaving no one outside to control it. Invoking The System has become a primary—perhaps the primary—modern means of totalizing and rationalizing our experience of the social and the political.

Dating what sounds like an attitude—when did systems first become something that *could* be blamed in this way?—would be a very difficult task. But if we link it to something that happened to a genre—to a specific change in mediation, in how system was used—then we can begin to understand "blaming the system" as a historical event. This event marked a new chapter in what I have termed the history of blame. Individual systems, as I have documented, had previously and frequently been blamed for exploding or falling short of the real. And we have also seen that the genre itself was looked on with suspicion by Bacon and others. But with system proliferating per my diagram of its linear rise—and proliferating into other genres and thus into the newly forming disciplines at the turn into the nineteenth century—it began to cast a shadow larger than itself.

In 1820, two decades after Malthus used system in the manner I have just described—embedding it in an *Essay*—William Godwin described a world that lived in the shadow of that act:

> It has not been enough attended to, how complete a revolution the Essay on Population proposes to effect in human affairs. Mr. Malthus is the most daring and gigantic of all innovators. (Godwin 1820, 622–623)

Since, as noted earlier, Godwin's own essay, "Of Avarice and Profusion" (1797), had been the immediate occasion of Malthus's effort, Godwin in this assessment was in part trying to keep up his end of the debate over perfectibility. But it was now twenty-two years later, and he is not so much arguing as bearing witness:

> Man, in the most dejected condition in which a human being can be placed, has still something within him which whispers him, "I belong to a world that is worth living in."
>
> Such was, and was admitted to be the state of the human species, previously to the appearance of the Essay on Population. Now let us see how, under the ascendancy of Mr. Malthus's theory, all this is completely reversed. (1820, 615–616)

Godwin's analysis of "how" primarily focuses on ideas, and they were certainly an important factor in Malthus's "revolution." But their meaning cannot fully explain why that revolution was experienced as "complete" and its instigator as "gigantic." What was, we might ask, the form of amplification?

Despite his focus on content, Godwin does point to that form as a matter of form, for his account of the ideas of the *Essay* attends to how those ideas were shaped into an argument. In Malthus's voice and then his own, Godwin proclaims to mankind:

> He [Malthus] has said, The evils of which you complain, do not lie within your reach to remove: they come from the laws of nature, and the unalterable impulse of human kind.
>
> But Mr. Malthus does not stop here. He presents us with a code of morality *conformable* to his creed. (621, emphasis mine)

What Godwin is ventriloquizing in the first paragraph of this quotation is Malthus's system—its principle acting as a law of nature, something that haunts us and yet is of us. But that system, the next paragraph tells us, did not stand still. It traveled, disciplining itself by striving for a fit with other kinds of knowledge—in this instance, moral knowledge.

This phenomenon—the traveling of embedded systems—was how, as Godwin saw it back then, "the Essay on Population … effect[ed]" its "revolution" in "human affairs." In retrospect, it is how knowledge of our own affairs came to be shaped by The System. Our characteristic reactions to it today—which define our sense of what it is—consist of various combinations of the responses Godwin inscribed. Well before he addressed the specific negatives of Malthus's revised code, Godwin observed, "The main and direct moral and lesson of the Essay on Population is passiveness" (616). That passivity could both carry "accents of despair" and erupt into the activity that Godwin himself pursued: blame. For him, and back then, the target could be personified as Malthus; for us, after being immersed for more than two centuries in writing configured in this manner, it is more amorphous, extending across all fields—not just population and morality—into The System as our particularly modern sense of things as they are.

THE POLITICS OF BLAME—THE NOVEL AND THE SYSTEM

Although great rivals in content—their debate over perfectibility helped to configure turn-of-the-century politics—Godwin and Malthus were comrades in form. In fact, that strategic similarity played an important role in elevating their debate to its prominent place in the war of ideas[4] at that time. They both found forms that amplified their arguments. Just as Malthus inserted his system of population into an accessible *Essay*, so Godwin made another form, the novel, the popular home of his "principles" of political justice. For Godwin, however, this embedding was not a matter of hurrying to join the conversation—per Malthus's apology—but an effort to keep his place in it.

The work that first earned him a spot in that conversation, *An Enquiry Concerning Political Justice* (1793), took a standard form of written system: a list of "principles" followed by expository prose. But the English reaction to the French Revolution and its aftermath made an exposition of that content in that form dangerous. With most of his friends in jail, Godwin took generic cover behind a fictitious narrative. Not only would a novel protect his person from his principles; as a genre growing in popularity, the novel would also, he hoped, appeal "to persons whom books of philosophy and science are never likely to reach" (3).

Attracting readers, however, was but the first step. As in Malthus, fitting system into another form enabled its principles to travel. Just as the principle of population carried consequences for the morality of those interpolated by its logic, so Godwin posited his political principles as reaching into "things passing in the moral world." "It is but of late," he insisted, "that the inestimable importance of political principles has been adequately comprehended" (3). Godwin set the vector for this reach into human behavior in his title, first invoking the existing system—*Things as They Are*—and then the fiction into which it traveled—*The Adventures of Caleb Williams* ([1794] 1988). The plot then elaborated a causal relationship between them.

Godwin's plotting was but one chapter in the interrelations between system and fiction—under such rubrics as "histories," "romances," and "novels"[5]—during the eighteenth century. At the head of chapter 2, for example, I cited Henry Fielding's narrator of *Tom Jones*: "I am not writing a System, but a History." That assertion of difference tells us that system was a form to be reckoned with in the eighteenth century. Fielding's fictions repeatedly feature such generic comparisons, each of them an effort to heighten the value of his own enterprise by juxtaposing it with other "high" forms. The preface to *Joseph Andrews* is the most elaborately plotted: Fielding tries to push his "comic romance" up the generic hierarchy by calling it a "comic Epic-Poem in Prose" predicated on a lost Homeric model (Fielding 1742, v). The point, of course, is to keep the right company, gaining value by proximity without being overshadowed.

In this case, Fielding appropriated the value of the epic as a prestigious form by invoking it as an absence—a move that corresponded to his audience's sense that the epic's power lay in the past,[6] whether classical or Miltonic. System, however, required a slightly different approach. Since its power as a genre, particularly in the wake of Newton, lay very much in the present, Fielding had to invoke system as a lively other. Thus, he valorized *Tom Jones* not as a reincarnation of a lost model but as a worthy alternative to a popular rival: system.

The interplay between the two throughout the eighteenth century included variations of two basic kinds: fiction-in-system and system-in-fiction. We have seen how the former emerged from the ongoing concern that the making of parts into wholes inevitably entailed acts of making things up. This was the temptation in Bacon's words, "to publish a dream of our imagination as a model of the world" (Bacon [1620] 2000, 24). Giving in to that temptation, as we saw Newton's editor, Roger Cotes, put it, meant "merely putting together a romance, elegant perhaps and charming, but nevertheless a romance" (Newton [1687] 1999, 386). This fear of fiction as the Achilles' heel of system was what led Adam Smith to characterize "all philosophical systems as mere inventions of the imagination, to connect together the otherwise disjointed and discordant phaenomena of nature" (Smith [1795] 1980, 105). What drove Smith and the other makers of Master Systems to forge ahead anyway was the conviction, inspired by the Newtonian system, that connection just might be possible: the world could be known. But they worried that in that final leap to completion—to matching the full complexity of nature—there would be a necessary fictiveness.

Systems undermined by faulty fictions became, in turn, a standard motif for writers of fictional narratives during the eighteenth century. In these examples of system-in-fiction, tales of system failure often configured both the portrayal of characters within novels and the overall shape of the novels themselves. Such protagonists, Eric Rothstein has argued, try

> to be led by a system. ... Thus Rasselas asks only one kind of question, whether each mode of life he sees can give him the same thing with which the Happy Valley has tantalized him, a static "choice of life" in which he can confidently repose. Toby Shandy, aching from and for war, interprets life as should an ideal soldier, and soldier in idea. The invalid Matthew Bramble makes nosology an absolute. (1975, 8)

All of these characters dramatize the fate of system portrayed as fictions relentlessly imposed on reality. And to the extent that the characters' behaviors dictate the overall movement of the plot, readers are forced to reenact the same pattern of disillusionment. In the case of *Rasselas*, for example, they experience, chapter by chapter, the insistent repetition of the Prince's systematic effort to order "life."

By the beginning of the nineteenth century, however, writers of fiction in the particular form of the novel articulated their relationship to system in a new, hierarchical fashion. "Let me make the novels of a country," wrote Anna Barbauld in 1810, "and let who will make the systems" (Barbauld 1810, I.61–62).[7] This bold assertion came at a crucial moment in the history of fiction and in the formation of the disciplinary category of Literature (Siskin 2001): the "institution" of the novel as we now know it (Brown 1997, 179–185). In fact, if Homer Brown is correct in linking that institutionalization to the editorial enterprise of Barbauld (*The British Novelists*) and Scott (*Ballantyne's Novelist's Library* 1821–1824),

Barbauld's statement is not only descriptive but constitutive. As the last sentence of the introductory essay that explains and justifies Barbauld's pioneering project of anthologizing the novel, it participated in what critics have called the rise of the novel.

To see that "rise"[8] in Brown's terms is to see it as part of the widespread historical phenomenon of institutionalization I have just described. From that perspective, Barbauld's downgrading of system was itself, ironically, a "systemization": the newly minted term for the procedure by which modern institutions were formed. Barbauld crowned the novel by putting it in systematic order. As we saw with "systems versus essays," the apparent rivalry of novels versus systems could and actually did work in a mutually productive fashion: more systems *and* more novels as exemplified by Godwin's two-for-one (a system-in-a-novel).

If Barbauld's claim announced the initial arrival of the novel as a high form in terms of its interactions with system, an article in the *Edinburgh Review* a generation later (1832) announced its maturity in the same manner ("Review of the *Waverly Novels* and *Tales of My Landlord*," 1832, 61–79). It singled out Sir Walter Scott for praise, but that praise formed the basis of a larger claim.[9] The knowledge that previously would have been contained in less entertaining genres was now, argued the *Review*, being conveyed in a more appealing form. The reform-minded public of 1832 wanted "facts," and fiction was now valued as a practical way of meeting the demand (Born 1995, 31–32).

"In consequence of this newly-enlarged view of the principles on which fiction should be written," declared the *Review*,

> we have, since the appearance of *Waverley*, seen the fruits of varied learning and experience displayed in that agreeable form; and we have even received from works of fiction what it would once have been thought preposterous to expect—information. … We have learnt, too, how greatly the sphere of the Novel may be extended, and how capable it is of becoming the vehicle almost of every species of popular knowledge. (77)

Using philosophical terms and mechanistic images that echo back through Newton and Bacon, the *Review* described the "extended" novel as beginning to behave as we have seen system behave: new "principles" allow it to be the "vehicle" for more things. In the terms I have suggested, the embedding of systems in other forms transformed their hosts—in this case, giving the novel a role in the work that system took up in the seventeenth and eighteenth centuries: the production and circulation of knowledge. To the surprise of the *Review*, novels began to function as what we would now call "information" systems.

Whereas Barbauld names system only to dismiss it, this reviewer does not directly reference any other genre, including system. The purpose here, after all, was to announce that the novel had come into its own. By this point—after more than a century of constant interplay—system had become a largely silent partner in celebrations of the novel. Those developmental details were part of the transformation of history I described in detail in chapter 2: the turn to disciplinary-specific histories of ideas animated by the agency of individual subjects. The tale of how system mediated the novel's rise requires both engaging system as a genre, and not just as an idea, and opting for a history geared to describing how genres mediate each other. Barbauld's statement makes sense only within such a history, as does Godwin's effort to embed system in the genre Barbauld helped to systematize less than two decades later.

The narrative of Caleb's "adventures" was configured by variations of the interplay with system I have described. It is presented as extending to include all things, for Caleb believes that only the total transparency of complete disclosure can defeat Falkland's obscure power. The narrative also features a variation of the practice of presenting each character as having a single informing principle—curiosity in Caleb's case, the "love of chivalry and romance" in Falkland's (Godwin 1988, 349, 6, 12).

The embedding of system in *Caleb Williams* put the concept of character itself into a new relationship to the "things" ordered by system: as Godwin observes in the Preface, they "intrude" into each other (3). As a result, Caleb comes to feel part of the very things that oppress him. This is, let me emphasize, a *formal* phenomenon in novels of that time—one that overrides distinctions of content, whether cast as dissenting versus conservative or as ideological versus aesthetic. Just consider how Jane Austen's independent heroines can find themselves only in marriage; how Walter Scott's Jacobite heroes inevitably tell the tale of the Union they opposed; and how the Union itself takes on the character of a "country," in Barbauld's terms, only as the product of its own collective fictions. They all experience, that is, that extraordinary sense of the social—of the totalizing of the social—that we now call "The System."

What makes that sense so powerful is the feedback loop between character and things: the more Caleb tries to "vindicate" his character, the more "things as they are" erase it—requiring, in turn, more vindication. This circularity of the novel of character is an effect of the embedding of system; it is another version of Kelley's thermostat talking to itself in an "endless internal bickering" about whether to turn the furnace on or off. Novels with embedded systems became self-regulating, their—if I may coin a term—logostats systemically keeping the internal conversation of character going. The result was a novel of regular selves whose internal bickering became the mark of the modern novel's fully rounded—think circular—character, their regularity valorized aesthetically as realism.

Godwin's goal in putting his system into fiction, however, extended beyond constructing realistic characters. He sought, in addition, to alter the reality of readers by including his audience in the feedback loop. Readers would thus become characters: "'I will write a tale, that shall constitute an epoch in the mind of the reader, that no one, after he has read it, shall ever be exactly the same man he was before.'"[10] To do so, he had to put the power of his system of *Political Justice* into the tale. Like any good systems engineer—think of the cloning of the IBM PC—Godwin proceeded "methodically," as he put it, to replicate the original through what we now call reverse engineering: "I began with my third volume, then proceeded to my second, and last of all grappled with the first" ([1794] 1988, 350).

The theory here is straightforward: to ensure that the narrative takes the characters, and the reader-as-character, to the desired state of "excitement" at the end, Godwin simply starts there, and then works back volume by volume. But here, again, the system embedded in the novel talks back: before publication, Godwin returned to the ending and rewrote it. Even in reverse, traveling proved difficult for system; as with Malthus, in the end it did not quite fit, requiring further revision. This novel told the tale I am retelling: how the genre of system became The System—and thus ushered in the modern era in the history of blame.

It does so by proceeding systematically; its comprehensive "consistency," we are told in the book's first paragraph, "is seldom attendant but upon truth" (5). Taking that connection between system and truth as a given, Caleb believes that anyone hearing his tale as he has configured it will necessarily believe him; the self-validating logic of system—the way it talks to itself—should make its truth self-evident. "Will you hear my justification?" he says to Mr. Collins. "I am as sure as I am of my existence that I can convince you of my purity" (320).

By the end, however, Caleb must face the fact that something has gone generically awry. Mr. Collins does not believe him; neither does Laura, nor, in the original ending, the judge. In that ending, his system, as he has constructed it, begins to break down, feature by feature. It cannot withstand Falkland's own systematic attack: he argues that the only principle that can explain all of the things that Caleb has recounted is "revenge" (342). Caleb cannot respond effectively, and he goes back to prison while Falkland flourishes.

Godwin portrays this defeat in a rather astonishing way—one that makes sense only within the history of system as a genre that I have been constructing. Back in prison and writing only in "short snatches," Caleb retreats from system to one of its primary eighteenth-century rivals: he struggles, Godwin specifies, "to proportion" an "essay" (344). For a true believer in system like Caleb, the psychological equivalent of this generic fall from complete system into fragmented essay is madness. The original manuscript ends with the snatches getting smaller and smaller until even the sentences break up. The exclamation points, dashes, and capital letters that remain work more like hieroglyphs than English prose, an appropriate marker for a man who becomes, in his own words, "an obelisk" (346).

Godwin may well be referring to this original ending in the final words of the published version. There, a very sane Caleb justifies completing—in perfect prose—his characterizations of himself and Falkland by noting that "the world may at least not hear and repeat a half-told and mangled tale." Whereas in the original ending, the collapse of system brings down the whole generic house, mangling the tale in which it is embedded, the novelistic structure survives in the revision. The fate of system then becomes an event *in* the novel's narrative and can thus be thematized. That's why the published version differs so radically from the original—the embedded system now formally gives this novel of character its abiding theme—or, as Godwin put it, the "new catastrophe."

At the very moment of vindication, with Falkland helpless before him, Caleb realizes that his system of vindication has worked, but too well—Falkland dies—and thus not well enough—Caleb has nothing to live for. "There must have been some dreadful mistake," he concludes, "in the train of argument that persuaded me to be the author of this hateful scene" (330). The thing Caleb's own "argument" missed, but the novel's now lucid ending conveys, is a fundamental incongruity, one articulated formally by the historically specific mix of novel and system, an incongruity between character and "things as they are":

> But of what use are talents and sentiments in the corrupt wilderness of human society? It is a rank and rotten soil, from which every finer shrub draws poison as it grows. All that in a happier field and a purer air would expand into virtue and germinate into usefulness, is thus converted into henbane and deadly nightshade. (336)

Embedded in the novel, system, I am arguing, becomes a vehicle not for rational explanation but for habitual blame. In the original ending, Falkland remains a tyrant and is clearly the reason for Caleb's descent into madness. Read the revision, and you have entered a different world—the world represented by and emerging from Barbauld's triumphant genre. It's the one we know, the one in which we know that it's no one's fault—and everyone's. For us, Caleb and Falkland are both victims of something other than each other, for we have The System to blame.

POLITICS AS BLAME—LIBERALISM AND THE SYSTEM

If we want to place the blame for having The System to blame, we need to look into forms of nonfiction as well as fiction, and a particularly important place to look is political economy. Through the mid- and late eighteenth century, it was a primary site for the totalizing and rationalizing of the social. By writing systems that presented society as functioning as a coherent System, Adam Smith and others laid the groundwork for turning it into an object that *could* be blamed. As we have seen, their project of methodizing knowledge met with considerable success, raising Scotland—and Britain's—profile in the learned world by flooding it with Master Systems that reordered what was known.

They did so not just out of nationalistic vanity—nor, as we have seen in recovering the importance of probability, a will to truth—but out the conviction that the increase of knowledge brought material gain, a link that was particularly active in political economy. In *The Wealth of Nations*, Smith systematized knowledge that first took shape, in the hands of William Petty and the Experimental Philosophy Club, as explicit efforts not only to discuss productivity but to be productive (Siskin 1988, 46–47). As David McNally points out, political economy "represented an attempt to theorize the inner dynamics of changes [toward capitalism] in order to shape and direct them" (McNally 1988, 1).

When the strategy of containing systems within a Master System gave way—under the pressure of its own success—to the dispersing of systems into other forms, the effect was twofold. Not only did political economy begin its nineteenth-century split into the narrow-but-deep specialties of political science and economics, but those newly specialized disciplines took on a new relationship to change. Whereas Smith's Master Systems were vehicles for the pursuit *of* change, embedded systems, I have been arguing, became objects *to* change—something to blame. Politics and economics thus turned to debates over reform—debates that saw system scale itself up by occupying every position in them.

This centrality of system in political behavior extended a thread that in Britain dates back to the Glorious Revolution. The social and political body that emerged at that time became known as "the English system," a polity characterized by a set of what were called "checks" through which king and Parliament kept each other's power in line. For much of the eighteenth century, the emphasis fell on the adjective, since this system was seen as constructed by and peculiar to the English. But here is where political economy came into play. Its proliferation in Master Systems and then embedded systems grounded that political system in economic norms, insisting that both "obey," as J. G. Merquior puts it, "the same explanatory principles and conform to the same regularities" (Merquior 1991, 25).

Having lost political sovereignty as a result of the Union earlier in the century and yet having gained economic flexibility through the facilitating of trade, Scotland was well served by this form of political economy. Its regular principles were not understood to be the product of human revolutions, Glorious or otherwise, but of the force that shaped them—shaped them through the natural workings of what came to be called an "invisible hand." Thus, for Scottish writers such as Adam Smith, "the English system" became but one example of what he called "the simple system of natural liberty"—a system, Smith insisted, that "establishes itself of its own accord" (27). As the system to which all historical and national variations tend, that "simple system" came to be known, simply, as "the system."

For Smith, the system was, by its nature *as* nature, blameless, its apparent problems not features of the system itself but of misguided efforts to change it through the imposition of other "systems either of preference or of restraint." In one of the earliest instances of system versus the system, a "Freeholder" published in 1784 *The source of the evil or, the system displayed*, a set of letters that announced itself on the title page as "printed and sold by all the booksellers in town and country."[11] It described the undermining of "the system of our mixed monarchy" by a "system of policy" instigated by "the Cabal" (Freeholder 1784, 10, 8, 19). This subject, the Freeholder tells his "Friends and Countrymen," is "next to what you owe to God, of the greatest importance to yourselves, and to your children" (3).

To make this case, the Freeholder first puts these systems into history:

> From the Revolution to the beginning of the present reign, the nation enjoyed, I may say, uninterruptedly the blessings secured to us at that glorious period. The different boundaries established by the Constitution, as then declared and ascertained, were, during that period of near eighty years, held sacred and inviolate. (6)

But "from the day that Mr. Pitt, the father, was first dismissed" (1761), the "secret abettors" of a new "pernicious system have been labouring to establish their power on the ruins of every maxim of constitutional Government." Rather than giving a "definition" of this system, the Freeholder opts to "trace it in its progress and effects, as the best means of giving you a complete knowledge of it in all of its parts and tendencies."

Characteristic features of the genre of system abound in this description, from the aspiration to completeness to the recognition of many parts. The stakes were political, and the argument was to resist change. But the mid-1780s was precisely the moment that the cyclical upturns and downturns that we now identify with nascent capitalism began (Foster 1974 19; Siskin 1998a, 140–142). By the 1790s, Britain was experiencing an extraordinary tension between wariness of large-scale change and a perceived need for it—a need fueled at that moment by both war with France and what David Fischer calls the "great wave" of inflation that brought the gap between prices and wages to a breaking point (Fischer 1996, 154). During the next quarter-century, change was demanded—and negotiated—in the growing economic as well as political discourse of institutional reform.

William Cobbett exemplified the ways in which system became a lingua franca of that discourse. "Where the 'English Jacobins' of the 1790s had proved vulnerable to the charge of systematic conspiracy," points out Kevin Gilmartin, "Cobbett turned the tables in launching a relentless attack on the systematic organization of British elites and the British government." His attack on system as a "totalizing mode of social organization" (Gilmartin 1997, para. 2)—a "thing" that should be blamed—contributed to the extraordinary shift in diction and institutions that reset British politics during the late eighteenth and early nineteenth centuries: Whig versus Tory metamorphosed into the opposition that marks political modernity in the West: Liberal versus Conservative.

The genre of system, I am arguing, played a crucial role in that metamorphosis: the increasingly common tactic on both sides of putting it into position to assume blame proved critical to the formation of liberalism as we know it. In blaming The System, we configure "things as they are" in a very particular way: as needing change, as capable of being changed, as providing the means of effecting that change, and, crucially, as *always* failing enough to maintain an ongoing need for change. That was how liberalism first took shape and how it still works to perpetuate itself: its object will always be in need of reform because those reforms will always fall short.

This cannot simply be understood as a complex of ideas thought up by individuals beforehand and then put into practice—insisting on ongoing failure, for example, would have been an unlikely strategy. This is, rather, "things as they are" reconfigured by the transformations of the genre of system that I have been describing. Liberalism emerged from those reconfigurations as a label for those who sought to mediate the demands for change. As its current association with big government suggests, its history is tied to totalizing conceptions of the social, particularly in regard to the bond between a society and its government. That bond was repeatedly tested at the turn of the century, resulting in the institution of the income tax in 1798, followed by governmental intervention in labor relations through the Combination Acts of 1799 and 1800.

System's role in those debates and in the ones that followed, just as with its role in the instituting of the novel, emerged in its interplay with other genres. In these policy matters, that chemistry entailed the various kinds—including the oral and the journalistic—that gave form to reform. Both the political debate over the Reform Bill, as well as the contemporaneous economic debate over the Poor Law, were shaped by those interactions. Just as system configured character in the novel, those debates came to be haunted by the legacy of Master Systems: the manner in which their totalizing drive was dispersed, feature by feature, into the newly mixed genres of the political and economic sciences.

In the famous speech by Thomas Macaulay that electrified Parliament into passing the Reform Bill, for example, references to system mixed with apostrophic turns to the House, geographical description, brief historical narratives, and features of many other genres, but there was no attempt to master all of the parts within a single whole. Thus, at the key moment, when Macaulay juxtaposed the demand for interventionist change with an invocation of the system that insists on the principle of hands off, he slipped the knot, insisting not on a systematic resolution but on conducting the argument in other terms. The System, as he embedded it, was no longer an encompassing form for methodizing and arranging everything, but a feature to be negotiated.

"It is said that the system works well," Macaulay asserted. "I deny it." This sounds as if it must have been a systematic claim about the truth of the System, but the mixing I have described allowed another startling alternative. Listen to this rhetorical turn:

> The House of Commons is, in the language of Mr. Burke, a check, not on the people, but for the people. While that check is efficient there is no reason to fear that the King or the nobles will oppress the people. But if that check requires checking, how is it to be checked? If the salt shall lose its savour, wherewith shall we season it? The distrust with which the nation regard this House may be unjust. But what then? Can you remove that distrust? That it exists cannot be denied. (Macauley [1831] 1965, 59–60)

This is the mix of modern politics. The systematic question, "Is the system really flawed?" *simply doesn't matter*—maybe yes, maybe no, but it's time to act. You do not need to judge the validity of the system; you can simply blame it. Embedded within the discourse of reform as transformed by Macaulay, The System preserved itself by becoming an object in the history of blame. In Macaulay's famous words announcing the end of the systematic opposition between interventionist change and laissez faire and the start of modern, liberal democracy, "Reform, that you may preserve."

The economic side of the reform agenda may have looked different, but the final object at issue was exactly the same. As Macaulay's philosophical rhetoric turned the House politically, the Poor Law debate turned to official statistics; the new language of economics thus highlighted and expanded the role that numbers had played in political economy since Petty's exercises in political arithmetic. But the same commissions that gathered the figures also sought verbal testimony about the need for change, and, there, as in the Reform Bill debate, a now-familiar object dominated the proceedings. Asked about the need to replace the old law with the new one, Joseph Ellison of Dewsbury exclaimed, "What a pity that a system that has worked so well, and has produced so much good, should be now broken up!" Despite his protests, the powers that be did just that, leaving Mr. Ellison to reenact the gesture that was rapidly becoming a habit on all sides of every issue: as the poor rates ceased and the workhouses opened, he blamed The System—in this case, for changing the system.

TAKING THE BLAME—THE HANDS OF WOMEN

In this maelstrom of systems undoing systems, victims of all kinds proliferated across the political spectrum, aggravating existing fault lines not only of politics and wealth but also of gender. We have had particular difficulty grasping what happened along that last divide, because we tend to conflate liberalism's rights of man with the rights of woman and assume that they mutually supported each other. But as Anthony Arblaster points out, "the issue [of women] has a uniquely ironic relation to liberalism" for a politics that

> takes as its focus the individual and the rights and freedom of the individual, ought logically to make no distinction between persons on the ground of sex. ... It might therefore be thought that the oppression of women ... would be particularly obnoxious to liberals. Yet liberals who championed women's rights, such as John Stuart Mill ... stand out by virtue of their isolation. Their consistency was seen at the time as eccentricity. (Arblaster 1984, 232)

Arblaster himself stands out from other twentieth-century students of liberalism, particularly men, in his even mentioning gender in his history. There are, of course, notable exceptions, particularly scholarship by women on issues of contract, but even those efforts do not explain the earlier silence cited by Arblaster.

To do that, we must start with what was present—the norm that made Mill eccentric. No other author was more normal in the discourse of liberalism then and still today than Adam Smith—not in the system-specific way I have been elaborating but in the history of ideas. The bible of British liberalism in what is called its "negative" phase—individual freedom in the state as opposed to state intervention to ensure that freedom—was and is *The Wealth of Nations*. One of its foundational claims about human nature—like the "love of system" I highlighted earlier—first appeared in Smith's *The Theory of Moral Sentiments* (1759): "every man is by nature first and principally recommended to his own care."[12] The question of care is especially important when it comes to Smith, for the issue of who really takes care of what leads to the most famous passage in *The Wealth of Nations*.

Laski frames it as follows: it is man's

> good fortune that, as he attends to his own wants, he is "led by an invisible hand to promote an end which was no part of his intention." For Adam Smith the myriad spontaneous actions of individuals, made for their own private benefit, results, by a mysterious alchemy, in social good. (Laski [1936] 1977, 178)

Recent work such as Anne Mellor's *Mothers of the Nation* has gone a long way toward helping us to dispel the mystery of who actually transformed private good into public good during the early nineteenth century (Mellor 2000). But our purpose here is to focus on how Smith's alchemy worked within liberalism itself to absent women from its calculations. Liberalism as a politics that, as Laski and others have argued, has sought to articulate property's role in capitalism did not overlook women as a matter of accident or irony; it made them disappear—and in a manner that we can make visible in a history of blame.

Turning particular hands into an encompassing invisible hand was part and parcel of Smith's scaling up of all other systems into his own. But once that Master System became an object that did draw blame, its parts became targets: particular kinds of hands materialized as mysteriously as they had disappeared. The women whom political economy had left out of the discourse of rights and contracts—and that the Reform Bill left out of the vote—reappeared in this new political science of blame as fair game; they became the usual suspects of liberal reform so that the blame could be absorbed without really disturbing the fundamental norms of The System itself.

The targeting began in earnest in the mid-1830s, immediately after the passage of the bill. Attacks on individual women with the most obviously busy hands—as in Reverend Polwhele's sexing[13] of Mary Wollstonecraft (*The Unsex'd Females, a Poem* 1798)—gave way at that point to larger-scale broadsides. Instead of Polwhele's verse, these broadsides were launched in the mixed genres of the newly forming disciplines, particularly those we now call the social sciences. "Journalism and sociological discourse alike," points out Daniel Born, "buttressed the predominant belief that the sufferings of the poor could be traced to individual defective will or desire." With increasing frequency and precision during subsequent decades, these attacks narrowed in on their prey. "Prominent Victorian sociologists," Born argues, citing Anita Levy's work, "identified the ultimate cause of poverty and misery not as working conditions, hours, and pay, but rather the loose morals of poor *women*" (emphasis mine). In 1836, for example, Peter Gaskell identified the problem as "'sluttishness' versus wifely virtue" (Born 1995, 33).

Such pronouncements marked the emergence of modern liberalism. The century that began with the proliferation of system and of the rhetoric of reform came to a close as they mixed generically into the disciplinary liberalism of the modern state. That politics of blame and sluttishness has remained with us into the twenty-first century in attacks on feminism and on welfare mothers and in the fight for family values. My point in pointing out their persistence is not to participate in those politics but to document system's ongoing role in configuring them—in turning them into a norm of modernity that is reinforced every time we blame The System.

6 WRITING UPON SYSTEM—INSTITUTING CULTURE

> With Mr. Wordsworth and his friends, it is plain that their peculiarities of diction are things of choice, and not of accident. They write as they do upon principle and system.
> —Francis Jeffrey, 1807[1]

> If by "system" is meant—and this is the minimal sense of the word—a sort of consequence, coherence and insistence—a certain gathering together—there is an injunction to the system that I have never renounced, and never wished to. ... Deconstruction, without being anti-systematic, is on the contrary, and nevertheless, not only a search for, but itself a consequence of, the fact that the system is impossible.
> —Jacques Derrida, 1997[2]

PARTS INTO WHOLES—SYSTEM AND THE IMPERATIVE OF CULTURE

To put liberalism into the histories of mediation and blame—of its relations to the genre of system and that genre's effects—is to track its travels into categories and practices beyond its own immediate politics. The acting out of those politics of blame at the turn into the nineteenth century helped to configure a new public space—a space other than the overtly political—in which to adjudicate claims about what parts of The System worked and what did not, what was of lasting value and what needed to be reformed. Those deliberations became the domain of *culture*—a term first used in its modern sense in Britain in the late eighteenth and early nineteenth centuries. As a noun—rather than the verb deployed through most of the eighteenth century—*culture* arose at the same moment that system came into its own as an object to blame and liberalism was instituted as a platform for doing so.[3]

The evolution of the term from its earlier sense grounded in the agricultural was, crucially, twofold. First, on the basis of that earlier use as "a noun of process," specifically "the tending of natural growth" (Williams 1976, 87), culture played a crucial role in the natural and conjectural histories of the Enlightenment—narratives in which everything a society does contributes to its staged transition from nature into a product of the process of culture—that is, into the state of *being* a "culture." To see society in this way was to see it as a totality—one in which all of its behaviors, arts, beliefs, and institutions cohered into a meaningful whole.

Once *culture* entered our conceptual vocabulary, every group had one—in fact, every group still *has* to have one. Culture enacts the formal structure of system semantically as an imperative: parts (can and should) fit together into a whole. It entails—to pair it with the term that entered English at the same time—a habitual *systematization* of parts into wholes. We simply assume that culture is always already there.

"While cultural practices are thought to vary," observes David Radcliffe, "all literatures and societies are thought to behave as cultures." But if the enterprise of designating and studying cultures valorizes itself by valorizing difference—theoretical, political, historical, and literary—then why do we rely absolutely on a concept that, as Radcliffe puts it, "applies to all literatures and societies indifferently" (Radcliffe 1993, x–xi)? The answer lies in what first appears to be a surprising twist in the evolution of the term. At the same moment in the late eighteenth century that culture assumed its totalizing function, it also took on a second meaning. Reorganizing internally the very unity that it helped to contrive, culture also came to signify a subset of itself.[4] Among the totality of activities that we call a "culture," is "Culture": in Raymond Williams's words, "the works and practices of intellectual and especially artistic activity" (Williams 1976, 90) that we now valorize as special—both in quality and in constituting the specialized subject matter of particular disciplines.

This doubling is less surprising when we think of culture in its historical context of system and blame. Viewed in that way, we should expect culture to exhibit primary features of the totalizing form it enacts. Like system, it is scalable—both doubling down into a subset of itself and scaling up into increasingly comprehensive iterations: local cultures into a national culture, national cultures into cosmopolitan cultures, all cultures into human culture.

That doubling followed the logic of liberalism as culture took on the imperative to reform and improve the whole it comprises—to identify and foster, that is, the most cultured parts of culture. Culture thus became a celebrated battlefield in the history of blame, from Matthew Arnold railing against the philistines to the culture wars of the late twentieth century. System has been as central to those clashes of culture as it had been to Newton's struggles with philosophy and to the political debates over reform. And it thus helped to reconfigure knowledge into the threefold realm we now inhabit: humanities, natural sciences, and social sciences. But knowledge, per the grandmother's visit to school (chapters 4 and 5), was not all that was reconfigured; that "we" was also subject to system, and at every scale: from the collective instituting of its behaviors in new "cultural" forms down to what Mary Hays called "the proscribed little personage—I" (Hays 1798, 298), an entity whose origins we explore in chapter 7.

THE CULTURAL INSTITUTION OF LITERATURE

I turn to the particular cultural form called "Literature" not just to add another example of system's reach, but to extend as well our understanding of how that reach materialized. The systems of Enlightenment took shape in writing, but writing itself was also transformed: it was institutionalized. That process was in many ways complex, but the nature of the change was straightforward and captured quite clearly, as noted earlier, in the lexicographic record. The word that referred through most of the eighteenth century to all kinds of writing—*Britannica* still defined *literature* in the 1770s as simply "learning or skill in letters"—came, in the space of a few decades, to refer more narrowly to only certain texts within certain genres. Like high culture to Culture, Literature, if we think of it as now also featuring the upper-case, doubled down to become a subset of its formerly inclusive self, literature.

As courses in English Literature first appeared in the curriculum—and as English departments first appeared in schools two hundred years ago—Culture found in Literature a particularly special(ized) home (Siskin 2005, 1068–1069, 1080–1081). As it settled in, literary history became a story that was less about learning in general, per Bacon, and more about curating a discipline. Its interrelations with system precipitated various classificatory schemes, from genres to periods, but what came to occupy the center, as I argued in chapter 2, was the subject—in this case, the subject who wrote. The Author took up residence with culture as the primary causal agent of literary history. Authors brought Literature to life, populating anthologies and generating an ongoing supply of articles and books to delimit and celebrate the field.

It was in his essay "The Poetry of Pope," for example, that Thomas De Quincey enshrined the distinction of scale between literature and Literature. By 1848 he could simply state as a truism that the "definition" of *Literature* "is easily narrowed; and it is as easily expanded." What he then famously added was a binary of affect:

> There is, first, the literature of *knowledge*; and, secondly, the literature of *power*. The function of the first is—to *teach*; the function of the second is—to *move*: the first is a rudder; the second, an oar or a sail.[5]

To illustrate that distinction, he turned to Newton's *Principia* as "a book *militant* on earth from the first," drawing a crucial distinction between content and form:

> In all stages of its progress it would have to fight for its existence: 1st, as regards absolute truth; 2ndly, when that combat was over as regards its form or mode of presenting the truth. (14)

For De Quincey, the victory of Newton's content "destroy[ed]" his "book," for its truth "transmigrated into other forms."

Works of Literature, De Quincey asserted, as opposed to Newton's militant literary efforts, were works "triumphant for ever," and thus "never *can* transmigrate into new incarnations." But the possibility that De Quincey did not entertain—because his purpose was to valorize works of Literature as the unique creations of their Authors ("the Prometheus of Aeschylus")—was that content was not the only thing that "transmigrated into other forms"; *forms can transmigrate into other forms*. Systems travel. They traveled into other forms—and with the small- and large-scale effects I have been detailing. When viewed in the history of mediation, Newton's legacy has been much more than what he argued. How he argued—and in what—has continued to shape knowledge of all kinds, not just his own.

Literary history in its familiar, modern form has made its points by missing this point—missing it in both general (how forms travel and mix) and particular (system as very well-traveled) ways. To miss the form of system in particular is to miss out on its role in configuring culture—and thus culture's role in the political and social instituting of modernity. Recovering those roles requires putting them in histories that allow them to be seen—histories adequate to the forms as well as to the content of knowledge. That is the strategy and the ambition of this book. To that end, I turn in this chapter to Literature as an institutional form of culture. Since both culture and Literature took hold by doubling down into subsets of themselves, I will double up. What follows might best be described as a literary history of literary history.

This brief sample engages one period and one author, since those two features remain the organizational touchstones of modern literary history. My first stop is the period literary scholars call Romantic, but not just because our narrative has reached that point. The late eighteenth and early nineteenth centuries are a special stop for my formal focus, for it was the moment, as argued earlier, that the newly forming disciplines instituted themselves by telling their own histories.

Romanticism has been Literature's label for its own telling—for the historical tales literary study tells itself about the period in which Literature was constituted as a category and a discipline.[6] Not only were the first courses in English Literature taught and the first departments of English formed; the essay and the review—as well as the periodicals that contained them—assumed their modern forms at that same time. Together, those formal innovations erected the critical scaffolding needed to professionalize the field.

THE SECRET HISTORY OF ROMANTICISM—A SYNOPSIS

> Elle, qui mettait toute chose en système.
> ["She who brought everything to system."]
> —Rousseau speaking of his "Mama"

Gazing from that scaffold, students of Romanticism have been less likely to be on familiar terms with system than their Enlightenment counterparts. This has been in part due to our standard literary histories fixing on a specific set of terms and genres to signal and embody the spirit of that age: genius and imagination, fragment and lyric. A few references to system that at least sound Romantic in that conventional sense have surfaced, however, notably the "howl[ing]" battle cry of the poet William Blake's Los in *Jerusalem*:

> I must Create a System, or be enslav'd by another Mans
> I will not Reason & Compare: my business is to Create
> (Blake [1808] 1970, 151, 1.10.20–21)

It is hard, of course, to miss Blake at full rhetorical throttle, and thus in regard to system, he appears, as has often been the case, to be an exception—but an exception that can help to clarify the larger picture. System was as crucial to Romantic writing as Blake's embattled declaration makes it sound;[7] in fact, invocations of and accusations regarding system were the discursive weapons with which Romanticism configured itself.

Take, for example, the dividing up of writing into "schools." "Mr. Wordsworth is a System maker" bellowed the *Poetical Register* reviewer of *Poems in Two Volumes* (Woof 2001, 231), an attack famously echoed by the critic Francis Jeffrey in the quotation heading this chapter: "With Mr. Wordsworth and his friends, it is plain that their peculiarities of diction are things of choice, and not of accident. They write as they do upon principle and system." Grouped under the rubric of system, these writers became easy targets of withering attacks; in the very maneuver feared by Los, they were "enslav'd" by another man's system: the critical system of grouping under system. Taking Los's advice and, in defense, creating a system of one's own did not, of course, end such strife; rather, it provoked it. Thus, Leigh Hunt defended his "vulgarisms" to Lord Byron by asserting that "his style was a system," only to be dismissed systematically as a system maker. "When a man talks of system," wrote Byron, "his case is hopeless" (Marchand 1973–1994, VI.46). As case after case followed the principle articulated in Byron's *Don Juan*—"One system eats another up" (McGann 1980–93, V.559, XIV.2.5)—the discursive map of Britain was redrawn, with a "Lake School" of writers in the north of England centered around William Wordsworth now at one end and a "Cockney School" in London identified (derisively) with writers such as John Keats and Leigh Hunt at another.

Many other features of that map, from generations to genres, were similarly inscribed. In system, then, lies the secret history of

Romanticism. Perhaps it would be more accurate to call it the secreted history of Romanticism, since, as these quotations demonstrate, it was not a secret back then; it was made to disappear. When all behaviors, including the literary, are totalized, palimpsest is the order of the day. Thus, as we shall see in the next section, Matthew Arnold sought—under the banner of Culture—to overwrite Wordsworth's "system" with his "poetry."

When, however, we pull Romanticism out of Literature's history of ideas and authorship and put it into the history of mediation, system reappears—embedded and specialized—in the numbers we have already tracked for those decades. We can then rediscover system as so pervasive in the period that it effectively functioned as Romanticism's familiar, haunting it from its forbearers to its finish. Claims for Rousseau's paternal status as the father of Romanticism, for example, have been based on features supposedly inherited from his work, such as confession, emotion, and a focus on childhood. But when those confessions turn to the figure who becomes his surrogate mother, another form comes to the fore: he remembers his beloved "Mama" in the striking manner that heads this section: "She who brought everything to system" ["Elle, qui mettait toute chose en système"] (Rousseau [1782] 1891, i.238). His rebelliousness has been conventionally understood as a turn from social convention, but what he actually turned to was not the wilds of nature but system—something he found only in a cherished few, such as Mama and M. Salomon, "who spoke tolerably well on the system of the world" (241).

In fact, Rousseau plots the entire *Confessions* as a journey from system to system. That form is the focal point; his feeling subjects act on each other *through* systems:

> It is, therefore, at this period that I think I may fix the establishment of a system, since adopted by those by whom my fate has been determined, and which … will seem miraculous to persons who know not with what facility everything which favours the malignity of men is established. (163–164)

Even Rousseau's celebrated turn to himself is mediated through the secret spring of system. His first sense of himself as capable of "effecting a revolution" was his "system of music" (297), the fortunes of which take us from Book 6 through most of Book 7, and take him—in his dreams—back "to the feet" of Mama, "restored to herself" (290). When he walks "alone," he confesses, there is always one thing in his "pocket": "my grand system." For Rousseau, the writing of system was his primary connection to writing: a turn "back to literature … as a means of relieving my mind" (23).

David Hume's literary career and analytic practice were, as noted earlier, similarly informed by system. His skeptical centering of philosophical doubt did call "grand" systems into question, but only to enable system to do other kinds of work. Rather than turning from the Enlightenment dream of knowledge, he enabled new ways to realize it—ways that were naturalized in Romanticism and were his signal contributions to it. His most notorious bout of doubting, for example—the attack on miracles cited earlier—was occasioned by the redeployment of system that shaped his career: the shift from the self-contained systematizing of the *Treatise* to the dispersal of system into the open-ended "attempts" of essay and inquiry. *An Enquiry Concerning Human Understanding*, unconstrained by a Master System's imperative to totalize, could open up to illustration and digression, including protodisciplinary ventures into new areas of knowledge.

Hume thus added entirely new sections to those carried over from the *Treatise*, including one that pioneered the comparative history of religions—a comparison based on the principle that the purpose of a "miracle" "is to establish the particular system to which it is attributed."[8] Not only did this anticipate Blake's conjectural history of religion in *The Marriage of Heaven and Hell*—"Till a system was formed"—it elaborated a logic and vocabulary of system that linked Hume at his most doubtful with Los at his most defiant. Since every miracle is generative of a particular religious system, argued Hume, it has

> the same force, though more indirectly, to overthrow every other system. In destroying a rival system, it likewise destroys the credit of those miracles, on which that system was established; so that all the prodigies of different religions are to be regarded as contrary facts. (121–122)

This rhetorical resemblance to Blake, I must emphasize, is not the most critical connection to Romanticism. More consequential is what made it possible: system appearing in two complementary ways—redeployed as form and thematized as content. That combination then became, as we have seen, Godwin's generic strategy when he inserted his system of "things as they are" into the fictitious adventures of Caleb Williams.

To call such effects "Romantic" is also to recall our attention to the ways in which the period itself is such an effect. Here is Francis Jeffrey's initial attempt to order the writing of his time according to schools or sects. It appeared in the very first issue (1802) of the periodical designed to shift that genre's goal from comprehensive description to the mapping of critical difference—the *Edinburgh Review*:

> The author who is now before us [Southey], belongs to a *sect* of poets, that has established itself in this country within these ten or twelve years. ... The peculiar doctrines of this sect, it could not, perhaps, be very easy to explain; but, that they are *dissenters* from the established systems in poetry and criticism, is admitted, and proved indeed, by the whole tenor of their compositions. (Reiman 1972, II.415–425)[9]

Using systems as touchstones opened the way for a new kind of disciplinary knowledge, providing some our earliest "facts" about the period we call Romantic: a starting point of roughly 1790 for the initial break from what was "established," as well as a framework of "peculiar[ities]" on which differences could be mapped. The resulting map of sects/schools was then drawn in terms of the issue of writing "upon system."

To Jeffrey, adherence to systems suggested a lack of the gentlemanly virtue of disinterestedness. However, as the disciplinary distinction between the science and the humanities began to take hold during the Romantic period, the negatives metamorphosed. Within science, interested errors of deduction became lack of objectivity; within the humanities, interest, whether in school or party, became a challenge to a newly idealized subjectivity: the sincerity and thus authority of the creative individual.

Arguments about schools therefore were easily overwritten by the emphasis on individual writers that still preoccupies the period study of Literature—particularly the study of Romanticism as what has been called the "Big Six": Blake, Wordsworth, Coleridge, Byron, Shelley, and Keats. A period can map and remap its differences using many grids, and system was an ongoing, but increasingly invisible, parameter in more than one. In fact, it played a substantial role in drawing the other lines that have made Romanticism into a recognizable whole: generations, gender, and genre.

Schools and generations can be superimposed, of course—the first generation linked to the Lakes and the second identified either in opposition to the first or configured by internal rivalries, such as the attacks on Hunt's Cockney School. However, the first generation is also identified with a specific project, one that has been seen as Romantic in the grand scope of its ambition—Wordsworth compared it to a "Gothic church"—but, as we shall see in more detail in the next section, Enlightenment in its conceptual architecture. That church was the great "*Philosophical Poem*" that Coleridge demanded of Wordsworth. The charge to embed philosophy into poetry was not only a matter of content, but, Coleridge specified, a matter of form: Wordsworth was "to deliver upon authority a system of philosophy … in substance, what I have been all my life doing in my system of philosophy" (Coleridge 1971, II.177 entry 403).

One effect of embedding system into other forms was, as we have seen, to allow its principles to travel into new areas of inquiry. That was, in fact, precisely the metaphor that Wordsworth himself adopted in his prospectus to the project (1814): he "must tread on shadowy ground, must sink / Deep, and ascend aloft" in order to go philosophically where no man has gone before: "into" *man* ("Prospectus," 16–17, 1–2). Although the planned project, *The Recluse*, remained largely unwritten, Wordsworth was identified with this procedure, for purposes of praise or of blame, throughout his career. "Your poems," wrote John Wilson in 1802, "are of very great advantage to the world, from containing in them a system of philosophy" (Wordsworth and Wordsworth 1967, I.352–358).

Two decades later, however, that act of containment (including the travel metaphor) became the means by which generational difference was solidified:

> And Wordsworth, in a rather long Excursion
> (I think the quarto holds five hundred pages),
> Has given a sample from the vasty version
> Of his new system to perplex the sages.[10]

Byron did not publish these lines, but his discretion in not wishing to "attack the dog in the dark" did not save him from the intragenerational warfare that used the same kind of weapon. *Don Juan* was assailed not just for its "profanity" and "indecency," but for how they had been "embodied into the compactness of a system."[11]

Byron, however, was already used to such attacks, and three years earlier had momentarily stepped back from the fray to make an eerily prophetic act of periodization:

> With regard to poetry in general I am convinced the more I think of it—that he [Moore] and all—Scott—Southey—Wordsworth—Moore—Campbell—I—are all in the wrong—one as much as another—that we are upon a wrong revolutionary poetical system—or systems—not worth a damn in itself—& from which none but Rogers and Crabbe are free—and that the present & next generations will finally be of this opinion. (Marchand 1973–94,V.265)

These claims are now very familiar ones, for they are claims about a period—the period we now call Romantic—and they are familiar even to the extent of raising Crabbe as an exception.[12] What is less familiar is that the argument is cast in terms of "system."

One consequence of that casting—one that has now occasioned a recasting—is the exclusion of women writers from this period portrait. It is important to stress that this exclusion was not done after the fact—as an effect of labeling the period "Romantic" or of criticism under that label—though such activities have certainly reinforced it. This was exclusion done during the period of time in question. Many factors played a role, including the formation of a two-tier market, the rise of anthologies, the reshaping of the working lives of women, and the professionalization of criticism (Siskin 1998a, 210–227)—but the deployment of system as an analytic tool (often in the service of that new professionalism) clearly contributed. Arguing about system helped to construct a system of exclusion.

This does not mean that writing by women did not refer to system—Austen calls the engagement of Frank Churchill and Jane Fairfax a "system of secresy and concealment" (Austen [1815] 1998, 361)—nor does it mean that their writing was not configured by the disciplinary travels of system as a genre. In fact, Hays's slippage from system to outline is precisely such a configuration. Rather it means that the pervasiveness of system—through embedding and in the rhetoric of criticism—made it available as a tool with which to discriminate. That is why Anna Barbauld, perhaps trying to defend at least one part of the generic ground held by women during the period, introduced her collection of novels—as we saw earlier—by trying to appropriate some of system's power: "Let me make the novels of a country," she wrote in 1810, "and let who will make the systems."

The claim by the *Edinburgh Review* only two decades later ("Review of the *Waverly Novels* and *Tales of My Landlord*," 1832) that the novel had matured into a system-like source of information is yet another sign of how the period, configured itself through the use of system. That is, of course, still the standard end point for the period. But what kind of end? "Maturity" does not suggest either an absolute end or a radical change, but rather something that has stabilized, at least for the moment. Vast amounts of critical energy have been expended, however, under the very different assumption that an end must be a break of some kind—that we need to argue about whether Romanticism *stopped*.

To put this period into a history that tracks system's ongoing role in shaping knowledge—to do a literary history of our standard literary history—allows for a different picture. The period can then be seen to emerge from the late eighteenth-century proliferation of print—a proliferation that turned the Enlightenment production of knowledge in the direction of the disciplines. As Literature's label for that moment of transformation, Romanticism can be said to end but by no means stop when the destination was reached: when knowledge settled into disciplinarity and the modern modes of literary production became standard. By the 1830s, we have clear signals of the pace of change reaching a kind of watershed—a normalization: the processes of print were completely mechanized, the mass market was established, and Literature was a fully instituted form of culture (Siskin 1998a, 11–12). At that moment, the period became, in a sense, a system capable of talking to itself,[13] of sustaining the discipline's now familiar dialogues of the creative and the critical, and of high and low culture.

AESTHETICS, AUTHORSHIP, AND THE FATE OF SYSTEM

> His poetry is the reality, his philosophy—so far, at least, as it may put on the form and habit of "a scientific system of thought," and the more that it puts them on—is the illusion.
>
> —Matthew Arnold on William Wordsworth, 1879

My point in eavesdropping on Romanticism is to repeat its conversations in a different key—to hear and report them as a historical norm of modernity underwritten by system rather than as an ahistorical feature of Literature. The report would not be necessary, of course, if something had not happened to system. We need to recover an understanding of how its underwriting—the ways that it continued to shape knowledge—was overwritten by the instituting of culture. Literature's retrospective periodization of itself as Romantic contributed by leaving the ubiquity of system that I have just documented out of its portrayal of the "Spirit" of that age. To see that process in action—a process that stretched from the period's supposed close in the 1830s to the end of the century (Curran 2010, 2–3)—I turn to the second organizational touchstone of modern literary history: complementing period in Literature's self-description is its embodiment in Authors.[14]

William Wordsworth is our case study in part because his life was so long (1780–1850) that it extended both slightly before and long after the conventional dates of Romanticism. In following it, we can thus see how authors and periods shaped each other during the time frame in which writing was acculturated[15] into Literature. In this case, the sight is rather painful, for periodization worked on Wordsworth like a procrustean bed. To fit him to Romanticism—actually, as we shall see, he and the "–ism" mutually constituted each other—literary history sliced his career off at both ends.

Most famously, with aesthetic praise for a "Great Decade" (1798–1808) as the an(a)esthetic, the last forty-two very productive years were critically and expertly amputated. So too were earlier efforts judged to be too much in thrall of eighteenth-century features or proclivities, leaving a savory center cut for future consumption. The entire operation can be seen from our vantage point as a variation on systemization: the whole cut up into parts was then professionally reassembled and repurposed to tell the developmental tale of the modern subject (Siskin 1998a, 108).

Wordsworth holds a powerful place in literary history not just because he was treated this way, but because he helped to devise the treatment. Unlike the many writers whose claim to our attention lies in how we value their works, Wordsworth wrote works that participated in forming those values. During his long career, he helped to set the norms for modern critical practice—from the vocabulary of "creativity" and "imagination" to the procedure of close reading. Even the very category of English Literature itself, as the basis for a discipline with a national focus, owes much to his generic experiments and polemical interventions. His poetry and his prose brought that discipline to life as a social and cultural institution, animating it with a cast of characters and a purpose: solitary authors and sympathetic readers realizing themselves in acts of writing and reading.

None of these were, of course, single-handed efforts. But to acknowledge that Wordsworth's was but one voice, a part and product of his time, is only to reformulate the issue of singularity: why has that particular voice been so persistently singled out as representative—whether of "great" poetry, a time period (Romanticism), a place (nature), or, more recently, an "ideology?" William Hazlitt's formulation (1825, 231)—"Mr. Wordsworth's genius is a pure emanation of the Spirit of the Age"—seems initially to suggest an almost effortless passivity or even ventriloquism. But the next sentence pries person and period so thoroughly apart that "emanation" becomes a matter of luck. "Had he lived in any other period of the world," Hazlitt continues, "he would never have been heard of." This strange sundering raises more questions than it answers. What would the period have sounded like without him? Would we have heard of it? Its immediate effect, however, was to allow Hazlitt to reassert Wordsworth's "genius" as something distinctive and deliberate. It was shaped by a purpose or, to put it more precisely, a genre. Wordsworth's enterprise, Hazlitt emphasized, was "to compound a new system of poetry" (232). The emanation of the spirit of the age was system.

System, I am arguing, was the underlying reason that Wordsworth was heard of back then, and it remains an important reason that he continues to be heard. To say this is not to discount the appeal, aesthetic or otherwise, of any individual poem; rather, it acknowledges that what established that reputation from the start, amplifying it in ways that judgments of "taste" alone could not, was something else. Even John Stuart Mill, who credited Wordsworth with permanently curing his habitual depression, insisted that this was not a matter of the "intrinsic merits" of particular poems—"There have certainly been, even in our own age, greater poets than Wordsworth"—but of a "way of thinking": he became a Wordsworthian (148–149).

Readers could become Wordsworthians because there was a formal basis for that transformation: Wordsworth's "system" pervaded and penetrated all of his work. Matthew Arnold, who famously sought to save Wordsworth's "poetry" from the Wordsworthian "system" (see the quotation at the start of this section), concluded that effort by confessing that he too was the very thing that he wanted to be on "guard against." "I am a Wordsworthian myself," Arnold admitted, testifying to the seductive power of the system making he deplored (Wordsworth and Arnold 1879, xix, xxv). Although instituting Culture entailed, as I have been emphasizing, Authors as touchstones, Arnold was on guard even as he endorsed one. For Literature to enact the doubling of culture into its high form, he realized, the uppercase *L* had to be valorized over any individual name. At stake for the institution he helped to build and serve was the core value of the "poetry" itself; what he saw as the contingent business of philosophical system making had to be, in comparison, an "illusion."

The difficulty Arnold faced in asserting and maintaining that distinction was the clear and consistent shape of Wordsworth's oeuvre: how central and persistent a role the business of system played in it. Those plans materialized in a then familiar form of literary—in the inclusive sense of the term—ambition, a form familiar to Wordsworth and all of his contemporaries. By identifying Wordsworth with an effort "to compound a new system," Hazlitt was not only placing him in the "Age" of Revolution, but linking him to the project that made that age possible: the enterprise of system making that informed Enlightenment and had enabled Newton and Galileo.

Wordsworth's version of that enterprise was to extend Enlightenment into and through verse—to use poetry to make Enlightenment. At the very moment, 1784, that Kant was asking, "What Is Enlightenment?" (1970), the young Wordsworth was already trying his hand at using verse to pursue what Hume called the "human sciences"—the effort to do for human experience and behavior what Newton had done for the "world": turn it into knowledge by revealing its underlying "system" of rules. Enlightenment confidence that this neo-Newtonian project could be realized led by midcentury, as we have seen, to widespread experimentation with new objects and new tools. Human nature and physical nature were mined for more "things," particularly their extremes and what had not been engaged before—both the extremely common and the extremely unusual—and the examples subjected to new forms of written analysis.

In a gesture toward Galileo's and Newton's "language" of mathematics, for example, Edmund Burke's *A Philosophical Enquiry into the Origins of Our Ideas of the Sublime and the Beautiful* (1757) elaborated an entire system of human behavior by composing—and imposing—an exacting binary on all experiences: from viewing a precipice to playing marbles, they must fall under the rubrics of either the "sublime" or the "beautiful." Writers of the Gothic used the genres of romance and the novel toward the same end, deploying the supernatural to sort out issues of "probable," that is, "natural," human behavior. The contemporary poetry now called "aesthetic"—from Gray to Collins to the Graveyard School—had similar ambitions on the human. Following that term's original meaning—"science of the senses"—the aesthetic poets sought to affect readers in an almost mechanical fashion through sensations generated by specific types of imagery, including personifications of abstract qualities.

Tracking aesthetics from its origins as a Newtonian enterprise into a keyword for Culture can provide an important context for understanding Wordsworth's trajectory. The enterprise in this instance was to make sense of sense itself. In the mid-eighteenth century, Alexander Baumgarten and others divided the human into a higher faculty of reason, whose science is logic, versus a lower faculty of sensation, whose science was supposed to be aesthetics (Baumgarten 1750–1758). But what happened to that project in the early nineteenth century exemplified the palimpsestic power of Culture: the overwriting of the Newtonian project by the qualitative discourse of critical taste. This took time, and Newton's script was not easily effaced. As late as 1821, Samuel Coleridge, for example, was still negotiating the use of the rubric: "I wish I could find a more familiar word than aesthetic," he wrote, "for works of taste and criticism" (Coleridge 1995, LXXV, 938).

Coleridge's reluctance tells us two facts about histories of aesthetics that simply assume that those works have always been its work. First, they are what we might call Whig histories of Culture—tales that erase the temporality of that category's success. In them, we lose sight of how Newtonian projects turned into features of culture. Second, the Newtonian aesthetics that lost out—the science of sense—was not only different, but, by the early nineteenth century, had already begun to seem strange. That is why Coleridge was concerned with familiarity—of how to make the word itself "familiar" by plucking it from that project and applying it to work that had become mainstream. His comments were in a letter that sticks the label to one of the most popular general interest periodicals of his day: *Blackwood's Edinburgh Magazine*, he claims, is a "Philosophical, Philological, and Aesthetic Miscellany."

Blackwood's would not, however, have been miscellaneous enough to feature this kind of aesthetics:

> If you have tried how smooth globular bodies, as the marbles with which boys amuse themselves, have affected the touch when they are rolled backward and forward and over one another, you will easily conceive how sweetness ... affects the taste; for a single globe, (though somewhat pleasant to the feeling) yet by the regularity of its form, and the somewhat too sudden deviation of its parts from a right line, it is nothing near so pleasant to the touch as several globes, where the hand gently rises to one and falls to another; and this pleasure is greatly increased if the globes are in motion, and sliding over one another; for this soft variety prevents that weariness, which the uniform disposition of the several globes would otherwise produce. (Burke [1757/1759] 1958, 152–153)

In its exquisite attention to the physical senses—and in its attempt to stimulate and actually evoke them—this description by Edmund Burke is "aesthetic" in the etymological sense. The taste mentioned here is not judgmental but physiological: if it were not granulated, sugar could not taste sweet. The pleasure is of the body and not of a discriminatory power of mind. This is aesthetics before it was aestheticized.

For a brief period of time in the mid- and late eighteenth century, however, the different strains of aesthetics mixed with each other: the primary strategy for establishing a hierarchy of literary taste was, in fact, writing to the senses. This was the agenda explicitly theorized in 1757 by Burke's effort to elaborate an entire system of human nature and experience through the extension of the binary mentioned earlier: the absolute distinction between primary categories of pain (sublime) and pleasure (beauty). The detailed descriptions that extend that argument, such as the one I have just quoted, are so saturated with sense that they strike us now as a literalization of Burke's topic: "sub" ("up to") plus "lintel" ("limit"), that is, "over the top."

The poetry that followed this imperative of sense also quickly came to be seen as too much. Wordsworth's famous "problem" with Thomas Gray's sonnet in the Preface to *Lyrical Ballads*, for example, was that he determined that only six of its lines were "of any value" ([1798] 1965, 132, 134). The other eight, he asserted, are "curiously elaborate" in their "poetic diction" and are thus of no use in a "good poem." For Gray, however, those are exactly the lines that do serve a specific qualitative purpose: he heightened their diction in order to secure poetry's special place in an expanding—more genres and more texts—literary field. The personifications that Wordsworth condemned are deliberately centered and elaborated by Gray to realize an aesthetics of sense: by bringing abstract ideas and objects alive, they are supposed to become sources of heightened sensation. In other words, both writers shared the same goal of securing high literariness for poetry, but Wordsworth dismissed Gray for precisely the features that Gray employed to make that distinction.[16]

For Gray, that hierarchical distinction depended on maintaining a distinction of kind between poetry and everyday prose. Wordsworth dismissed his efforts because he was pushing an alternative: the collapsing of distinctions of kind into a new hierarchy of degree, aesthetic degree. Poetry, he insisted, is just like prose but more pleasurable. That is precisely the strategy that Coleridge invoked to justify his use of "aesthetic." "There is reason to hope," he argues,

> that the term 'aesthetic', will be brought into common use as soon as distinct thoughts and definite expressions shall once more become the requisite accomplishment of a gentleman … [and] above all, when from the self-evident truth, that what *in kind* constitutes the superiority of man to animal, the same *in degree* must constitute the superiority of men to each other. (emphasis mine)

As with Wordsworth, Coleridge's place in the literary history of Literature was amplified by his transitional role in the instituting of culture. On the one hand, both of their careers provide primary evidence for the ongoing centrality of system. On the other, both of those authors took on the disciplinary status of Authors by participating in the totalizing work of culture—and thus contributing, in the long run, to the overwriting of system's role in the shaping of that knowledge.

This relationship of culture to system is also confirmed by the fate of Burke's keyword. Writing before the aesthetics of taste eclipsed the Newtonian aesthetics of sense, Burke had no concept of "culture" to use in his *Enquiry*. Within a few decades of its midcentury publication, however, the use of his keyword, *sublime*, began to decline—decline, in fact, in inverse proportion to the growing use of the word *culture*.[17]

Starting his career while sublimity still held its sway, Wordsworth pursued many of its Newtonian strategies. His earliest efforts were almost programmatic attempts to sample those techniques: classifying experiences according to sensations, exaggerating the sensations for gothic effects, conveying the effects through a proliferation of personified images:

> But he, the stream's loud genius, seen
> The black arch'd boughs and rocks between
> That brood o'er one eternal night,
> Shoots from the cliff in robe of white.
> So oft in castle moated round
> In black damp dungeon underground,
> Strange forms are seen that, white and tall,
> Stand straight against the coal-black wall.
> (*Vale of Esthwaite*, lines 35–42)

Many of these techniques were abandoned, of course, and the rest reworked, but what is evident in such force in these early experiments is their epistemological confidence—the Newtonian and Enlightenment conviction that the full range of the human could be delineated and examined.

That historical purpose was the common ground that first helped to bring Wordsworth together with Coleridge in 1795 to plan their delivery of "a system of philosophy," a task that, as we saw earlier, Coleridge later simply referred to as "what I have been all my life doing." Wordsworth agreed to the same task, but to do it in a special way, demonstrating, in the process, why that way was special. Rather than just writing poetry informed by the systematizing concerns of Enlightenment, Wordsworth was to trump his predecessors by constructing an entire system in the form of a "great" poem.

The purpose of writing a complete system in a form different from its primary Enlightenment home of prose was twofold. First, embedding system in verse held out the prospect of extending Enlightenment principles into new areas of inquiry, allowing the system to explain new "things" by traveling. In fact, travel narrative was precisely the Enlightenment genre that Wordsworth himself adopted in his prospectus to the project: he "must tread on shadowy ground, must sink / Deep, and ascend aloft" in order to go philosophically where the Enlightenment had been working so hard to go: "into" man ("Prospectus," 16–17, 1–2). A new form also promised a second benefit: it might attract others into taking the journey. A poet, Wordsworth argued in the preface to his and Coleridge's first joint "experiment," *Lyrical Ballads,* was "a man speaking to men," and poetry, they hypothesized, might make that conversation more "pleasurable."

Although the great "Philosophical Poem," *The Recluse,* was never finished as planned, Wordsworth was identified with acts of system making, for purposes of praise or of blame, throughout his career. That is why Hazlitt used the genre of "system" to seal Wordsworth to the "Spirit of the Age"—a sealing that began with the reception of *Lyrical Ballads.* "Your poems," wrote John Wilson in 1802, "are of very great advantage to the world, from containing in them a system of philosophy." The label stuck: "Mr. Wordsworth is a System maker" bellowed the *Poetical Register* reviewer of *Poems in Two Volumes.*

The historical irony, then, of Wordsworth's canonical status as the representative Romantic was the Enlightenment nature of his undertaking. System, as I have been documenting, had been the Enlightenment's experimental crucible—its generic site—for attempting to mix parts into new wholes. Before engaging *Lyrical Ballads* as just such an experiment, we need to identify its conditions of possibility: what conditions and contexts brought Wordsworth to this intersection of "system" and "poetry," and what was at stake in mixing them?

The condition that enabled and shaped Wordsworth's career as a writer was revolution—but not only the political revolution that erupted in France. Britain itself witnessed a technological revolution during Wordsworth's lifetime. He was born the year the eraser was invented (1770) and died the year the *New York Times* was founded (1850). During those eighty years, Britain became a society fully saturated by the practices and products of that technology. As we have seen, print had proliferated only slowly and unevenly throughout most of the eighteenth century; not until the final two decades did publication rates take off, rising dramatically and consistently into the next century. Wordsworth and print grew up together.

They also matured together. The level of technological and social change connected to print's proliferation remained high—and highly innovative—through the first two decades of the nineteenth century. But by the 1830s, as we just saw in addressing the "end" of Romanticism, those innovations consolidated into a new norm that formed the foundation of the modern world of print. The basic printing processes were fully mechanized; a mass market, including "high" and "low" levels of "culture," was formed; and the flow of writing itself, from production through consumption, was regulated and rationalized by new forms of knowledge and work. Courses in English Literature, for example, began to be taught in the universities, and a section of a school could, in Ian Michael's words, "be known as the English Department" (Michael 1987, 377).

His voice amplified by his reputation as a system maker, Wordsworth played a key role in the making of these new social systems. As part of that historical process, Wordsworth's Enlightenment efforts were naturalized. Terms that were for him terms of classification—such as *imagination* and *fancy*, and the differences between them—took on new lives of their own. What had been philosophical principles offered to explain the "human" in Newtonian terms became the psychological truths of the humanities. Ian Reid has documented in detail how the institutionalization of English literary studies emerged from Wordsworth's ways of thinking and writing and has "continued to take its bearings from them" (Reid 2004, x). Thus, even when not directly reading Wordsworth's works, we encounter them—"appropriated and mediated"—in the courses and syllabi and the anthologies and periodicals that shape our experience of the literary as Literature.

Wordsworth's own initial experiences of the literary were shaped by matters that he and his sister, Dorothy, called in their letters "pecuniary." The growing market for print meant more competition as well as opportunity, and for the Wordsworths, this became an issue of genre: poetry versus prose. Even after publishing *An Evening Walk* and *Descriptive Sketches* in 1793, Wordsworth spent spring 1794 planning to begin his literary career in earnest by starting a "monthly miscellany" on moral issues "from which some emolument might be drawn." Although this purposeful focus on issues that mattered seemed to promise an audience, engaging them in the highly competitive periodical marketplace was clearly risky.

The stage was thus set for Coleridge's arrival and an innovative reworking of form and content: bring purpose to poetry by making poetry do the work of Enlightenment. For the Scots, as we have seen, that work was to make the achievements of English philosophy newly accessible, "arranging and methodizing their discoveries," as Adam Smith put it, in the genre of system and in the "simple and natural" language extolled by Hume. Wordsworth and Coleridge took up that task, retaining both the genre and the style but claiming a distinctive role in that undertaking by turning from prose to poetry.

The burden of their first joint project, then, was to justify that turn: can poetry do it, and do it better than prose? The 1798 edition of *Lyrical Ballads, With a Few Other Poems*[18] thus began with an Advertisement that formulated the case for verse noted earlier into two complementary principles. First, poetry was up to the task of Enlightenment—of philosophical inquiry into Man—because its "honourable characteristic" is that "its materials are to be found in every subject which can interest the human mind." Poetry conceived in this capacious manner could thus compete with the standard prose form of Enlightenment: the "books of moral philosophy" that Wordsworth pointedly derides in two of the lyrical ballads ("Expostulation and Reply" and "The Tables Turned"). Wordsworth offered his "book" of verse as a cure for those others, claiming that it "contains a natural delineation of human passions, human characters, and human incidents."

The second principle was that poetry's delineations were more "natural" than those of prose because poetry teaches, as "Nature" does, through pleasure. As in a modern advertisement, Wordsworth captured and focused attention on a single, salient difference: poetry's competitive advantage over prose, he argued, is that it offers more pleasure, and thus can better engage more readers. But how do you measure pleasure? Wordsworth's Newtonian solution was to offer the poems themselves as "experiments" for gauging how much better poetry could be.

To stage that experiment, Wordsworth and Coleridge brought together the two basic components of a "philosophical" system: explanatory principles and objects to be explained. They then ran the experiment to see if the system worked. *Lyrical Ballads* was, in this sense, a pilot project for the great philosophical poem. Its readers willingly played their roles in this Enlightenment enterprise. As I noted earlier, all of the reviewers of the 1798 edition read the poems through the frame of the Advertisement, and Wordsworth was explicitly portrayed as writing "upon" system.

Through all of those efforts, Wordsworth's explanations of what is "valuable" in poetry and of the nature of poetic "language" remained remarkably consistent with the system project—and consistently controversial. "Poems to which any value can be attached," he argued in the Preface, had to be products not just of poetic "sensibility" but also of "thought." As such, they were distinguished by seriousness of "purpose"—that purpose being both engaging a "wide variety of subjects" to "discover what is really important to men" and "enlighten[ing] the "understanding of the reader."

That combination of new subject matter and new ways to connect to the reader set the scene for modern poetry. As mad mothers, idiot boys, and sailors entered poems as characters or dramatic narrators, the poet required new techniques to elicit the reader's sympathetic participation in this accumulating of things for system. Already motivated by the competitive marketplace of print culture, Wordsworth had been working through all of his early works to claim a kinship with an audience for print that was growing—and changing—along with the technology. In both "An Evening Walk" and "Descriptive Sketches," he frequently left the text to reach toward the reader in self-justifying notes and dedications. In "Salisbury Plain," he violated the "general rule by which narrative pieces ought to be governed" in an effort to engage "reader's sympathies." In the play *The Borderers*, stage considerations were similarly ignored in the hope "that the reader might be moved."

When Wordsworth and Coleridge started their joint project in 1797, their conversations turned, Coleridge tells us, on exactly the same topic: "the power of exciting the sympathy of the reader." The suturing of system and sympathy was a task, they decided, requiring a division of labor. Coleridge's projected role for *Lyrical Ballads* was to focus on the supernatural, procuring for it "that willing suspension of disbelief for the moment, which constitutes poetic faith," while Wordsworth aimed to explore the "every day" to awaken "the mind's attention from the lethargy of custom." The single object of both endeavors was to make readers susceptible to a new system.

All of these gestures, however, are informed by a strategy that has proven to be one of Wordsworth's most powerful legacies:[19] he contributed to the early nineteenth-century phenomenon of systemization by systematizing the field of literature in terms of kind and degree. The prose-versus-poetry argument provided a template. By rejecting inherited distinctions of kind between poetry and prose—such as the use of an elite poetical language, particularly abstract personifications—Wordsworth leveled the playing field. But that was just the first of a two-part maneuver. He collapsed differences of kind in order to institute differences in degree: poetry is more pleasurable than prose.

Wordsworth's description of the poet followed the same sequence. He collapsed distinctions of kind—a poet "is a man speaking to men"—only to insist on differences in degree: the poet has a "more lively sensibility, more enthusiasm and tenderness." Pairing after pairing went through the same mill—speech and writing, writer and reader. In every case, one type of hierarchy supplanted another. This move is of a piece, of course, with the contemporaneous political formation of liberal democracies—all men are created equal, but then they improve themselves by degree. In fact, it played a crucial role in that formation by configuring literary activity into the field of culture I have been describing—a field that underwrote the formation of the modern nation-state and provided a rationale (cultural superiority) for the empire building that began in earnest for Britain during Wordsworth's lifetime.

For Britain, that culture was, and is, middle-class culture—and here too Wordsworth's recasting of poetry played a role. The controversy, initiated by Coleridge, over whether Wordsworth really used the "language really used by men" has largely obscured this contribution. The purpose of this argument, like so many of the others cited, was to construct an audience not just for poetry but—and this is where the standard literary histories always miss out—for a *system embedded in poetry*. Like Newton before him, trying to figure out what kinds of system would work with kinds of readers, Wordsworth revised and retargeted. Both faced audiences of varying knowledge and expertise. The "language" problem for

Newton was mathematics; he revised Book 3 to use less and then more of it. Wordsworth negotiated the differences by trying to establish a new common ground. On the one hand, he rejected the language of elitists who "confer honour upon themselves." On the other, he sought to purify the language of "humble and rustic life" from what is "disgust[ing]" in that "rank in society." Taking just the features he desired from those above and from those below, and rejecting the rest, he constituted the "middle" as not only a viable but preferable space. Occupying it did not require agreement over the specifics: the newly forming class now had a linguistic/cultural arena in which it could articulate itself in the very act of disagreement—and in relationship to Wordsworth's system.

Articulating his own relationship to system was Wordsworth's other pilot project for the great philosophical poem. What was at stake in writing *The Prelude* can easily be lost in its forest of biographical detail. For modern literary history and criticism, after all, the biographical turn is habitual: the more we think we know about an author's life, the greater the temptation to use that knowledge to explain the work. So much scholarship on Wordsworth and Romanticism, as well as on literature in general, does just that; we "adjudicate," as David Simpson (1993) has put it, "by reference to a biographical entity." To do so, of course, is to follow Wordsworth's own lead—even if for the wrong reasons—confirming yet another mark that he made on modernity. Most obviously in the thousands of lines of *The Prelude*, Wordsworth helped to pioneer the epistemological centering of the individual self, as well as the constituting of that self out of detailed, developmental narratives.

"Lives" had, of course, been written before, but not in this particular way.[20] To take egotism as the central issue in Wordsworth's decision to write about himself at a length—in his own words, "unprecedented in Literary history" (Wordsworth and Wordsworth 1967, 586–587)—is to miss the historical point: without a developmental strategy for understanding and representing that life, the most egotistical person in the world would not—in fact, could not—write that much about himself. But once he did, the writing of others changed. The fixation on all of those private details and the fascination with possible developmental crises (in Wordsworth criticism, for example, William's relationship with his sister, Dorothy, and the effect of his brother John's death) still constitute a surprisingly large share of critical attention. His reputation owes a great deal to that type of attention, both directly—by sustaining interest in him, his "circle," and his work—and indirectly—by testifying to the ongoing power of a primary effect of that work: the centering of the private, psychologized self as a primary object of knowledge.

Here again, however, we confront the historical irony behind my including Romanticism and Wordsworth in this book: by identifying this centering of the newly detailed "I" with that period and that author, we lose sight of its Enlightenment underpinnings. To borrow a concept from Newton's mechanics, the momentum of Enlightenment helped to carry the self toward the center. Wordsworth took his life not as somehow valuable in itself but as the readily available experimental ground for pursuing that neo-Newtonian effort to identify the laws of "man"—to systematize the human. The genre now called "autobiography" gave him the "things" for his "principles" to explain; it provided the raw materials for his system. Like Malthus, he understood the power of his system to be enhanced by accumulating more things, whether that meant finding those things in the lower classes, as Malthus did, or in his own life.

That focus on system was why Wordsworth went to what now seem like puzzling lengths to police the use of the life he detailed. He persistently refused, for example, to organize his poems in order of composition, objecting adamantly to any personal chronology "placing interest in himself above interest in the subjects treated by him." Every variation of the classification systems that he experimented with throughout his career gave priority to the general and the social—on what was or could be shared: faculties ("Fancy"), stages ("Childhood"), issues ("Liberty and Order"), genres ("Sonnets"), places ("Scotland"). These categories cast the poems as parts of a whole—and, for Wordsworth, the primary whole was not himself-as-author but the system-in-verse that was his lifelong project.

To understand that project, then, we need to watch what Wordsworth made of the life, not just what he told us about it. The French Revolution, for example, certainly changed his life. In recounting it, Wordsworth even provided the types of intimate details—including a veiled reference to his illegitimate French daughter—that can tease one out of critical thought and into a voyeuristic discourse of personality. How he placed the Revolution in *The Prelude,* however, reclassifies it as more than just a moment of personal or social crisis; it emerges as an enabling event in the history of the project. The project, after all, is what *The Prelude* is prelude to. Wordsworth called it "the Poem to Coleridge," for its purpose was to affirm his willingness to take on the task of system his friend had proposed. It was, in other words, a long, self-explanatory "yes."

Historians of Literature, however, have read Wordsworth's despair over the course of events in France—from "Bliss was in that dawn to be alive" to "all things tending fast / To deprivation" (*Prelude 1850*, XI.108, 253–254)—as a rejection not only of the Revolution but of the Enlightenment itself. Cast as an argument about literary periods, this becomes the familiar formula of Romanticism as a transcendent reaction against all that came before. Wordsworth, however, describes the arc of this "crisis" (XI.306) not as a turning away from but as a return to what he was and had been doing. Recovery is portrayed as an act of remembering, and thus of continuity: "I was no further changed / Than as a clouded and a waning moon" (XI.343–344). He was, he discovered, "still / A Poet" (XI.346–347), one who "found / Once more in Man an object of delight" (XIII.48–49).

What remained unchanged, in other words, was his commitment to the Enlightenment project of system making that was peculiarly his: the task of "making verse / Deal boldly with substantial things." Echoing Bacon's call for a new science of "things as they are," Wordsworth reaffirmed poetry as his vehicle for writing a system of "men as they are men" (XIII.226–235). The wrong turns of the French Revolution returned him to "right reason," to the effort to discover the "steadfast laws" of human nature—laws as "fixed" as those that Newton himself had found in the natural movements of the "waning moon" (XIII.22–23.372).

Wordsworth's strategy in *The Prelude,* then, was to use both the natural world and the unnatural worldliness of the Revolution to shape his "life" into an argument for the system project. What Wordsworth called the first "Retrospect" shone the light of the Enlightenment on the new object of inquiry: "Love of Nature Leading to Love of Man" (title of Book VIII). The Revolution revisited and then re-presented that object in terms of a distinction between "external man" and "men within themselves." For Wordsworth, the Revolution in retrospect was a giant arrow flashing in one direction: "in" (XIII.226–228, 49). It signaled that the "mystery of man" was a matter of "depth" (XII.272).

Earlier writers had, of course, gestured in that direction—from some version of the external toward some version of the internal—but Wordsworth linked depth with development. Together with writers in other genres, such as Jane Austen in the novel, Wordsworth helped to distinguish the modern self from earlier versions—it values itself as deep because depth testifies to an ongoing capacity to change. Living in a society in which a growing economy was altering the lives of a growing population increasingly concerned with the phenomenon of growing up, these writers changed change itself into something natural, one of "those fixed laws" of "Man." Development, while admitting change as substantial and inevitable, also robs it of its sting by converting it into evidence of a more important continuity—of the persistence of the self over time.

That was a major worry of Enlightenment empirical philosophy, one that Wordsworth directly addressed in "Ode: Intimations of Immortality from Recollections of Early Childhood." Whatever we make of its metaphysical trimmings, that poem's basic structure is four stanzas posing the problem of change—"Wither is fled the visionary gleam?"—and seven stanzas offering developmental depth as the solution. The rewarding residue is "Thoughts that do often lie too deep for tears."

Austen's *Emma* followed suit, domesticating the changes challenging its heroine and Highbury by following the logic of the "Ode": an inward turn plots developmental change as depth of self—"To understand, thoroughly understand her own heart was the first endeavor." Between the two-part *Prelude* of 1799 and the thirteen-book *Prelude* of 1805, Wordsworth replotted his tale of self to effect the same turn. What had been a straightforward chronology of experiences linked to specific spots (e.g., the Derwent River, the Vale) was twisted back on itself, confounding time and space into life as a system that could talk to itself. Most famously, to recount the French Revolution in the manner I have just described—as a return to self and the system project rather than a rejection of the past—Wordsworth lifted early memories from the start of the narrative and placed them in Book XI (XII in 1850) as "spots of time" (258). There, they layer memories—of the incident itself, of remembering the incident, of remembering that remembering—and thus evoke change as a guarantee of continuity: the "mystery" of "depth" in which we experience the connection of parts to the whole.

The Prelude was not only Wordsworth's venue for giving value to depth, but of claiming it for himself—and for a very specific purpose. His refusal to publish it during his lifetime provides a clue as to what he was after. Doesn't everyone have a piece of writing that they never publish in final form, at times allow to circulate, and are always quick to revise? At its most effective, it attempts to expand autobiography to epic dimensions by recounting supposedly heroic acts.

The Prelude is, quite simply, the most extraordinary résumé in British literary history. Wordsworth referred to it as a "review" of what "had qualified him for such employment," the "such" referring to the system project: "the task of my life." The vitae-like detailing of personal identity turned maturation into a preoccupation with occupation: in this case, how to become, behave like, and perform as a professional poet *and* system maker—someone who can always turn parts into wholes. Such status and coherence is described in a now-familiar sequence in which personal background (the influence of Nature), formal education (Cambridge and books), and significant life experiences (the French Revolution) all contribute to a concluding epiphany of professional purpose: "what we have loved, / Others will love, and we will teach them how" (Book XIV.448–449).

Wordsworth's prophecy became the logic of modern literary study—of English as a pedagogical discipline defined by a "love" of Literature. That was his institutional legacy. But falling in love—with Literature or anything else—requires a sense of depth into which one can fall, and that is what led to Wordsworth's formal legacy. He mined older forms of verse for features capable of delivering that effect. I venture very briefly into some examples not as a departure from my focus on system into the literary, but as a way of showing how Wordsworth's centering of system helped to configure literariness, the quality that distinguished Literature from literature.

"Lines Composed a Few Miles above Tintern Abbey" was an experiment performed on the ode, specifically on that form's characteristic feature of abrupt transitions. The transitions in this poem are not only abrupt but repeatedly cast in the negative: "If this / Be but a vain belief, yet, oh!"; "Nor perchance, / If I were not thus taught." When introduced in this fashion, each verse paragraph is not just a part of the whole poem. Rather, each one is itself potentially a whole (i.e., "this is what I owe to 'Tintern Abbey'"), that, when juxtaposed negatively with the others, both reinterprets them ("that may not be what I really learned") and is itself reinterpreted ("this may not be either"). The abrupt reversals produce the effect of spontaneity, and as one transition supersedes an other, the repeated formulations suggest depth and therefore new heights. The poem thus enacts and amplifies the message for which Wordsworth is still known: the claim that there is, as he put it in *The Prelude*, "something evermore about to be" (VI.608). But what is less known—what has been overwritten—was that the guarantee of "more" was also an affirmation of form, of the scalability of system, of wholes always becoming parts of larger wholes.

"Tintern Abbey," however, turned out to be an experimental dead end for Wordsworth and others, its ode-like turns yielding few imitations. The sonnet proved to be a much more productive model of a form talking to itself. Wordsworth was fascinated by its internal transitions, particularly the division in the "strict Italian model" between octet and sestet. He saw that break as enacting the developmental ideal of change within continuity, the disruption deep within the body of the sonnet suturing the parts into a whole. So enamored was he of the sonnet "as a piece of architecture" that he embedded sonnets and sonnet parts (fourteen-, eight-, and six-line units) into other forms; he even split one book of the thirteen-book *Prelude* of 1805 into two, yielding a fourteen-part whole. Capitalizing on how such wholes could in turn become parts of larger wholes, Wordsworth also turned in earnest to sonnet sequences during the latter part of his career. All of these experiments in a kind of Newtonian mathematics of prosody helped to transform the sonnet into a basic formal building block for nineteenth- and twentieth-century poetry, analogous to the couplet in the eighteenth century.

Recovering this contribution would be one benefit of loosening the stranglehold that Arnold's Great Decade argument—his slicing up of the career into ten years (1798–1808) of true "poetry" that needed to be saved from the poet's system making—has had on our understanding of Wordsworth. In all of his decades, Wordsworth continued to bring system's ambition—its determination to include more things—to poetry, as in the *Sonnets Dedicated to Liberty and Order* (1835–1845) and the *Sonnets upon the Punishment of Death* (1841). This was an ambition conveyed not only in ideas but through form. In his Newtonian drive to assemble the laws of Man within a single system—his great philosophical poem—Wordsworth made every individual poem do double duty. Each one was supposed to "stand upon its own merits" *and* be part of the larger whole. Works working on these multiple levels added to the effect of depth. Readers encountering experimental forms, advertisements, prefaces, afterwords, classification systems, and a broad range of subject matter began to experience poetry as a genre that was now, in a new way, imposingly deep and capable of conveying information.

Many of them were impressed but very wary of being imposed upon—thus Francis Jeffrey's famous denunciation of *The Excursion* (1814): "This will never do." Jeffrey admired Wordsworth but was worried about a poet having a system—worried because he knew that the system was amplifying Wordsworth's voice, giving it a depth of authority even among those who resented it. It was not that he wanted that voice silenced—precisely the opposite: he wanted to take over the job of amplifying it by filtering out system. Like Arnold after him, Jeffrey was working to establish the position of the cultural critic as a necessary mediator of Literature.

The fate of Wordsworth's system, then, need not be reduced to a tale of personal failure per the "Great Decade" argument. It can be told more productively as a subplot of the emergence of the modern disciplines from Enlightenment through the embedding of system. By 1814, Wordsworth was himself contributing to that enterprise and the accompanying emergence of culture: he narrowed the scope of his great "philosophical poem" to "the sensations and opinions of a poet living in retirement." Announcing that it was not his "intention formally to announce a system," he turned the task of "extracting the system" over to the "Reader" (Wordsworth [1814] 1974, III.6).

From that moment in the early nineteenth century to the turn into the twenty-first—from Jeffrey's denunciation of system to Derrida's valedictory refusal to "renounce" it[21] (see the quotations heading this chapter)—Newton's choice of guesswork has shaped the experience of Literature for both authors and readers alike. Embedded in other forms, system underwrote the narrowing of knowledge into disciplines and continues to haunt the borders between them, reminding us that knowledge was once shaped differently. A literary history of literary history confirms that the instituting of modernity has been another manifestation of system's ongoing power—and Literature, as the institutional empowerment of writing, presents a particularly telling example. Without Literature, culture would not have become so central to modernity, and, without system, writing would not have assumed the stature of Literature.

7 SECRETLY SEEKING SYSTEM—INSTITUTING THE SOCIAL

> But you know the number of times I've wished that I had never heard of the damned word [culture]. I have become more aware of its difficulties, not less, as I have gone on.
> —Raymond Williams, 1979[1]

> But you cannot behold him till he be reveald in his System.
> —William Blake, 1804–c. 1820[2]

SUBJECTS FOR SYSTEMS—"THE PROSCRIBED LITTLE PERSONAGE—I"

One of the best witnesses we have had to putting the category of culture into history—as I sought to do in the last chapter—was Raymond Williams. Not only did his work in *Keywords* (1976) help to establish culture's emergence in the late eighteenth century; he also, as he revealed in the interview quoted above, bore witness to the ongoing and growing power of the term. So vexed was he about this twist in his legacy—that the interviewer would credit him with popularizing rather than historicizing "culture"—that he felt compelled to renounce the term and act on that renunciation. The sequel to *Culture and Society* (1958) received a new name: the book originally titled *Essays and Principles in the Theory of Culture* became *The Long Revolution* (1961). Under similar scrutiny, "media" became as ideologically suspect as "culture," its use a "blockage," Williams decided, to our understanding of what he considered his "key move": the realization that the means of communication are themselves a material means—a historically specific means—of social production (1979, 133, 139).

Directing attention to the act of mediation itself and its effects—rather than *the* media—was one of the primary reasons that William Warner and I introduced the notion of a "history of mediation."[3] Within that history, we can examine forms of communication and behavior—as well as those who use and exhibit them—not as features of culture but as forms with effects, including the formation of "culture" itself and the human parts that come to constitute that whole. My focus in this chapter, then, is on system and self.

To provide a sense of what is at stake in mapping the intersection between them, let us turn briefly back to Mary Hays's struggle with the form of her *Appeal*:

> Though I have not certainly, the vanity to believe myself equal to the task of fulfilling the title of this chapter ["what women ought to be"], to the entire satisfaction of my readers; nor what is perhaps in the first instance still more discouraging, even to that of my own; yet having once adopted I shall retain it, as it expresses exactly what I wish to accomplish, however I may fail in the execution. (1798, 125)

In just this one sentence, the first-person pronoun *I* appears four times. Hays herself was clearly surprised by what expressively filled the gap between the authority of system and her "free" outline. She concluded her appeal with six pages of "apology for a fault which is perhaps too obvious to escape notice":

> It must be confessed, that, 'the monosyllable' alleged to be 'dear to authors'—that the proscribed little personage—I—unfortunately occurs, remarkably often, in the foregoing pages. (298)

What is most remarkable to us today—after so many "I"s in so much writing—is that such an extended apology would have seemed necessary. Hays speculates that vanity might be at work—and adds, in one last satiric jab, that men may be less susceptible—but if we scale up to include genre as well as gender, there is another explanation.

Whereas what Hays called a "regular" system is self-contained, a system embedded as a part within another form cannot account fully for the whole. To put this in Kevin Kelly's (1994) thermostatic terms, when systems are extended through disciplinary travel so that they can no longer talk to themselves, another kind of self must be formally interpolated to do the talking. Although admittedly incomplete, Hays's system, by her own account, still "expresses exactly what I wish to accomplish." It does so, I would argue, precisely because it authorized that newly expressive "I." Her statement is performative. The "I" it invokes is the modern subject that we identified in chapter 2 as the personification of agency, the deeply expressive self conventionally linked to Romantic lyricism. I connect it here to system first rather than to the lyric because specific changes I have identified in the use of system—such as embeddedness and the inclusion of more "things"—were the conditions of possibility for changing the subject. Hays's "I" became the subject in both senses of the term: it is both the subject that expresses *and* the content of that expression. The "I" that literally proliferates in Hays's prose is what she was appealing to men to recognize.

Those men were themselves busy with their own embedded systems and thus their own "I"s. Wordsworth, as we have just seen, found that the more he attempted to embed a philosophical system in verse, the more he had to apologize for writing about himself. The result was incompletion on a very large scale; his big "I" was almost up to eight thousand lines when he died without publishing it. Hays's "monosyllable" had also proved irresistible to William Godwin. He started *Caleb Williams* (1974)—a system embedded in fictitious adventure—in the third-person, but felt compelled, he claimed, to switch to the first.

There does appear to be a formal pattern here—a pattern that emerged from particular uses of particular forms. What has been famously psychologized in the history of ideas as the narcissism of modernity—its centering of self—manifests differently in the histories of mediation and blame. Narrow-but-deep selves (the "little personage" now authorized as "exactly" "express[ive]") emerged from system's role in mediating the formation of narrow-but-deep disciplines. Cast in these terms, this connection should not surprise—nor should our sense that in blaming The System we are also somehow blaming ourselves.

The legacy of our system-filled Enlightenment, then, has been not only an object that invites blame—society totalized as The System—but also an ongoing supply of subjects authorized to oblige. When that subject began to take shape, however, its relationship to system was not dominated by blame. To recover what came before and is now coming after, we need to look backward and forward to the modes of selfhood that overlapped with Hays's little personage back then and are mixing with it today. In this chapter we will juxtapose one of the most distinctive social features of the eighteenth century—the proliferation of clubs, conferences, and other associational forms—with efforts to meet the new social obligations of the twenty-first century: the need to relate to the computing machines that now populate our algorithmically configured landscape.

Although we conventionally apply the term only to our current situation, the issue in both cases was and is finding a working interface with a dominant technology. In the eighteenth century, that meant the technology of writing,[4] and now it means the technology of computing. Most of today's interface tales, featuring mice, fingers, and windows, take it for granted that the goal is to alter the machine to make it comfortable for the user. To the extent we don't feel that way—that we feel *we* are being made to change—we blame it on the system: both The System in general and the particular computing system on which we are working. By putting system into a history of mediation, however, we can recover alternative scenarios—including one in which system was an object of desire for those desiring change.

THE "CHARACTER OF MEMBERS"—CLUBS AS INTERFACES

My strategy here is to tell two tales in tandem—one from the eighteenth century and one from the twentieth—in the hope that doing so will trigger the effect that William Warner and I call "digital retroaction"—a feedback loop between past and present that transforms our experience of both. On the one hand, the current shift into computation provides new concepts, tools, and evidence to bring to bear on the past, altering what we find and how we engage it. On the other hand, our altered histories speak back to the present, inflecting our sense of the futures that might emerge from it. This is, then, an encapsulated version of one of the strategies shaping this entire book.

Here are the protagonists of the two tales: a report on *Augmenting Human Intellect: A Conceptual Framework* by a systems engineer named Douglas Engelbart, from 1962, and *An Account of the Fair Intellectual-Club of Edinburgh, 1720*. Some readers may recognize Engelbart's name and realize the importance of the report to our desktops, but for now, let us focus only on the representations on the page in figure 7.1 and then tell

AN

ACCOUNT

OF THE

Fair Intellectual-Club

In EDINBURGH:

IN A

LETTER

To a Honourable Member

OF AN

ATHENIAN SOCIETY there.

By a young LADY, the SECRETARY of the CLUB.

NATURE, *who gave, till She could give no more,*
On WOMAN *lavish'd all Her precious Store;*
Who now courts solid and substantial Praise,
Nor values Beauty, wedded to a Face:
Her MIND *peculiar Ornaments desires,*
And Vertues proper to her SEX *requires.*

ROW's CALL.

Enter'd in Stationer's-Hall.

EDINBURGH,
Printed by J. M'EUEN and COMPANY, and to be sold at the said *J. M'Euen's* Shop in *Edinburgh*, and T. Cox at the *Amsterdam* Coffee-House near the Royal Exchange in *London*, 1720. Price One Shilling.

Title page of an *Account* (1720) and diagram from a *Report* (1962). The *Report* courtesy of SRI International.

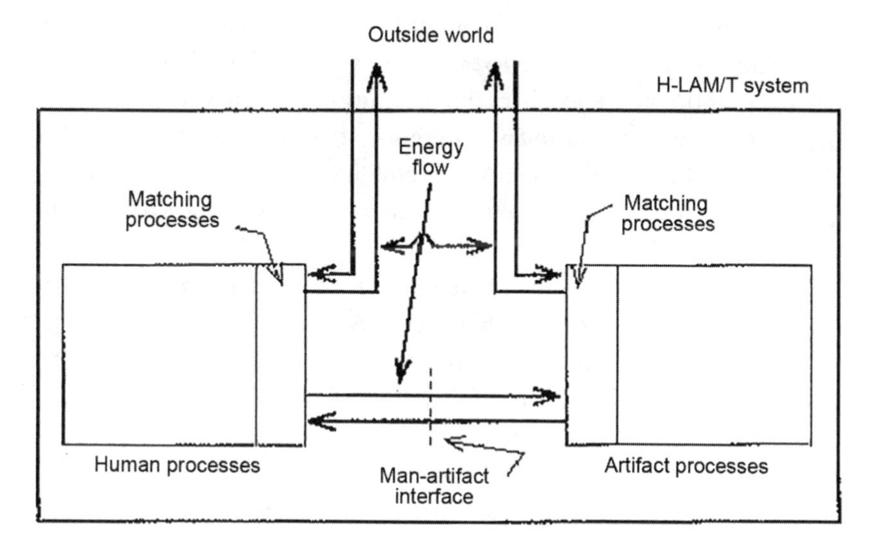

Figure 7.1

the tales. Neither of these representations looks very human—no conventional faces here—but that is because both are meant to enact the volatility of the category: the place where forms of technology and human processes meet to transform each other into something other.

Both the diagram and the title page are mixed forms that communicate through design—shapes and lines and fonts—and through words, words that invoke a further multiplicity of forms: Engelbart's hybrid form H-LAM/T (Human using Language, Artifacts, Methodology, in which he is Trained) and the eighteenth-century text's combination of genres—account, letter, poetry—a combination that, as we shall see, serves not just to describe but to constitute the Fair Intellectuals. On the one hand, the title page of the *Account* can be seen as a diagram, its lines illustrating the flow between the outside world (the "Athenian Society") and the human ("a young Lady") inside the club. On the other, Engelbart's diagram can be read as an account of social interaction between what's outside (the "world") and what's inside the box (the "human" clubbing with the "artifact"). Engelbart, as noted on the diagram, insists that his protagonist is a "system," and, I will now argue, that is precisely what the club aspired to be.

To understand what these women were aspiring to, we need some sense of where they came from. On the climb up the Royal Mile to its haunting Castle, veer to the left as you cross the wide esplanade leading to the entrance and look over the stone wall. Down below you will see a magnificent turreted building—often mistaken for Hogwarts—whose foundation stone was laid in 1628. Heriot's Hospital is a hospital in the old sense of the word—a charitable institution often devoted to education. It remains a school today—a school of choice for the children of Edinburgh professionals. Next to the building, you will see some lovely gardens. That is where the account begins a few years before Adam Smith was born.

The account was actually a confession; it revealed a secret—one that was briefly exposed back then but then was reconcealed in the archives. I found it bound deeply into the oblivion of a miscellany in the Special Collections of the University of Glasgow Library. The title page alone tells a strange tale. In 1720 in a print shop in Edinburgh and a coffeehouse in London, an unusual document was put up for sale. "An Account of the Fair Intellectual-Club, &c.,"[5] signed "M. C." and dated

July 28, 1719, purports to be the first admission in print of the existence of a small secret society of women formed in 1717 in Edinburgh. Betrayed by one of their own to a gentleman friend, the group felt compelled to protect itself against untoward speculations by clarifying its history and purpose.

The emphasis is on the social; these women came together to work jointly toward a specific goal. The word they used was the same one that Engelbart uses again and again to explain his project: *improvement*. These women, however, gathered not just for individual improvement; rather, they convened for the "Improvement of one another"—a concept repeated throughout their text. This mutual improvement was necessary, they argued, because of the "Disadvantages that our Sex in General … labour under, for want of an established Order and Method in our Conversation" (6). Entrance requirements were strikingly practical: "at her first Admission," the new member must deliver ten shillings for the poor and "entertain the Club with a written Harangue" (8). Reading and writing as the bases for inducing proper conversation were their core activities. With a membership limited to nine, like the muses, they took particular interest in the sister arts of poetry, music, and painting.

In addition to the history of the formation of the group, the text includes a description of the "Rules and Constitutions" and transcriptions of talks given by two different members. One of the most striking moments comes in rule XV, which defines the conditions for terminating membership. The two that are specifically named make for a particularly telling pair: you left the club through either "Death" or "Marriage" (9). These are young women—the age of admission was between fifteen and twenty, with only a brief window for intellectual activity before their "mutual Love and Friendship" for each other gave way to men or to some other deadening "occurrence" in the "Course of Providence." Not only is this—to my knowledge—one of the first, if not the first, club of this type on record at that time in Britain; the use of the term *intellectual* to describe the club's purpose was one of the earliest uses of that word in its particularly modern sense: a word designating a very specific set of personal and social behaviors—behaviors to which these individuals aspired.

Their aspiration was transformation, and the technology they deployed was writing—by which I mean again, following Raymond Williams, the interrelated practices of writing, silent reading, and—inevitably, as they discovered when their secret leaked—print. The club's explicit purpose was to rewrite its members—its secretary, author of this account, emphasizes that the three founders all insisted on a "written scheme" (5). What they in fact wrote down was the imperative to write everything down: "All the *Speeches, Poems, Pictures, &c.* done by any Member … are carefully kept" (25). Even the oral dimensions of club behavior were grounded in writing: the "harangues" I noted earlier had to be "written" (8).

But why? Why did these young women want to spend their brief moment of freedom between childhood and men immersed in writing? To what end did they write? What did they want out of writing? What did they think writing could do? This is, of course, a strange set of questions for an academic to ask. On the other hand, is there an academic who hasn't posed similar ones to herself? Perhaps as a thought experiment, then, we might for the moment remove all human selves from the formulation and opt for a Richard Dawkins–like reversal. In his infamous "selfish gene" conceit, human beings are the environment that genes render in order to survive; *we* are the means to *their* end.[6] In Dawkins's scenario, they, not us, are the main players in the Darwinian drama of survival of the fittest.

With Engelbart in mind, computer science can also help us think through these issues of causation. A "Theory of Conversation" has become a centerpiece of that discipline's efforts at interface design. To make software work, computer scientists found they had to ask not only "How does the user tell the software what to do?" but also, in a Dawkins-like reversal, "How does the software tell the user what it can do?"

So here is the reformulation for the *Account*: why did writing gather humans into groups? One answer, preposterous as it may first seem, was to propagate itself: this was writing's way to ensure its own ongoing proliferation. Why groups? Because writing had to solve the same kind of reproductive problem faced by the most virulent of the rogue genes we call viruses: what to do when the host population becomes so bedridden that the virus has no means or place to go—when success, that is, threatens the organism with extinction. I am thinking here of Ebola, the terrible disease that consumes its victims from the internal membranes out. We have no cure, but, in many outbreaks in less populated and isolated areas the disease has helped to contain itself by killing its hosts so quickly and efficiently that they cannot spread it very far.[7]

Think of a human infected by the writing bug; she spends, as we all know, more and more time alone in chairs (if not in bed), all forms of contact disrupted by the isolating experiences of immersion in a book or dissertation or fixation on a blank sheet of paper. For writing to spread—to circulate extensively and efficiently through entire populations—that tendency must be counteracted by new forms of sociability and publicity. In eighteenth-century Britain, the most obvious example was pornography; as the newly public form of private desire, it helped to incite that century's rise of writing—a role that it appears to be reprising for the new technologies of the Web.

A less obvious but more pervasive behavior was what David Kaufer and Kathleen Carley call "reverse vicariousness." They use the term to describe how infectious contact with writing was initially and then repeatedly secured—even over the spatial, temporal, and social distances that writing itself opened up (Kaufer and Carley 1993, 12–13). Knowledge of the text need not, they point out, come through the text itself; it can cross the gap

> through reviews, and through word of mouth. … Even nonreaders can positively register at social gatherings that they "know of" the book without actually having seen or read it first hand. We might call this phenomenon *reverse vicariousness,* because we normally think of immediate viewing or reading as vicarious experiences for [the] face-to-face interaction [the text describes]. But, in this case, a viewer or reader uses face-to-face interaction [with others] to experience the viewer or reader role vicariously. (66)

Writing, if you'll indulge the "selfish gene" conceit a bit longer, puts *us* into new forms of face-to-face interaction in order to maximize its own circulation. The very technology that threatens to isolate us preserves itself by reinvoking the social through the workings of reverse vicariousness. Thus, the more saturated we are by writing, reading, and print, the higher the premium we put on an expanding repertoire of face-to-face encounters, from improvement clubs to tutorials to cocktail parties.

To see those forms as the invention of writing may sound strange, but doing so allows us to demystify them, particularly the assumption that face-to-face interaction is "real life." For the Fair Intellectuals, meeting in person meant meeting in writing. Really meeting meant meeting in this mediated way. This is where two of the histories I have been deploying intersect. The history of mediation points to writing as a primary form of mediation at the time these women met. And the history of the real tells us that the large-scale effect of that form of mediation was to alter the experience of the real.[8] Starting in the eighteenth century, writing, through reverse vicariousness, made the face-to-face "real"—a connection it made so effectively that in the early twenty-first century, it is still common to be suspicious of anything else, including today's new technologies of distance communication (e.g., social media, the Web, videoconferencing, virtual reality): we assume them to be less real and thus in some way dangerous (Siskin 2001a).[9]

Scholarship on eighteenth-century clubs—and on the public sphere in general—has certainly shared this predilection for the face-to-face and the civility we associate with it.[10] As we shall see, what the members of the Fair Intellectual-Club saw when they were face-to-face was *not* each other's faces, and their behavior was far from civil. Their practice highlights the shortcuts taken by so many efforts at historical inquiry into clubs. In the *TLS* review of Peter Clark's important book, *British Clubs and Societies* (2000), for example, the reviewer simply dispenses with Clark's arguments to assert that clubbing must be "innate in the British character." He then posits a cause-and-effect sequence that is not in the book: "English clubs were founded by individuals for no other reason than the wish to be in each other's company" and "inevitably, bureaucracy soon followed" (Mitchell 2000, 10).

Both Clark and *An Account of the Fair Intellectual-Club* tell a different story, one in which, I will show, what we think of as bureaucratic systematizing is anything but an afterthought. By focusing on why Britain's associational world stood out from the rest of Europe, Clark sets the stage for the more specific concerns raised by Edinburgh's Fair Intellectuals of 1717: Why Scotland? Why so early? Why women?

I have already pulled two key terms from the account of the Club: *Improvement* and *writing*. As Davis McElroy has shown, an explicit "impulse toward improvement" in Scotland can be dated well back into the seventeenth century when the Scots Parliament passed legislation "Encouraging Trade and Manufactures" (McElroy 1969, 1). The economic interest was always paired with a moral one, for the Kirk (the Church of Scotland) understood its mission to be ongoing reformation. Fueled by the parallel economic divides between Highlands and Lowlands, and between Scotland and England, as well as the fervor of the church, twin efforts to improve manufactures and morals proliferated.

Those efforts assumed the social form of clubs and societies as the issue of the Union came to a head at the turn of the century. It did so in writing and through a variety of genres, from the reports of church commissions to the record books of the Reformation societies to the new periodicals such as Defoe's *Review*. As the colonial fiasco in Darien—Scotland's failed attempt to establish a trading outpost in the New World—further debilitated the Scottish economy, the debate over the Union became, explicitly, a debate over improvement. With help from "English Farmers, English Graziers, and English Husbandmen," Defoe wrote, Scottish agriculture would flourish (see McElroy 1969, 6).

When the Act of Union passed in 1707 and Scotland lost its own official state apparatus, the work of improvement fell more fully on church-sponsored and voluntary groupings. Clubbing became a cultural phenomenon in the modern sense of the term, for the act differentiated the political and economic, now linked to England, from what remained under Scottish control: the legal, religious, and educational—those areas, that is, having to do with the passing down, regulation, and valorization of distinctive traits, customs, knowledge, and beliefs. In other words, Scotland was left with the parts that were then totalized into what we would now call culture.[11]

The technology that built that particular infrastructure was writing, and writing could be said to require, as I have just suggested, groups of humans in order to work. Thus, to find them gathering together in 1717 for the purpose of improvement is not surprising, especially after the failure of the first Jacobite rebellion two years earlier confirmed the new configuration of the Union. What is startling is to find so early in the century young women doing it on their own and in secret—startling, that is, until we put the Fair Intellectual-Club and its members' predilection for writing into the history of mediation.

Sadie Plant's (1997) daring and still largely isolated effort to zoom out to a Raymond Williams–like "long revolution" of the relationship of women to technology can help us to put that predilection into perspective. Texts, as Plant notes, are abbreviated textiles in the etymological sense and in the material sense of the woven strands on which we write (69). Her primary argument in *Zeros and Ones* is that women have been at the secret center of technological change for a long time. Women "gathered at one another's houses," she points out, "to spin, sew, weave, and have fellowship. Spinning yarns, fabricating fictions, fashioning fashions ...: the textures of woven cloth functioned as means of communication and information storage long *before* anything was written down" (65, emphasis mine).

The same picture emerges, argues Plant, as we move toward the present—to the start of the electronic era of alternatives to needing to write things down. Once again we find women forming into groups at sites of technological change. In the switching rooms of the first telephone systems, the women operators summoned into groups by the new electronic technology to spin its wires formed informal networks within the formal ones they were paid to maintain. "In several [telephone] exchanges," notes Plant, "reading clubs were formed, in others flower and vegetable gardens, and a women's athletic club in another" (123).

To encounter the Fair Intellectual-Club as an episode in this historical sequence is to gain a new and powerful sense of women's relationship to writing. The turn to that technology—its centrality and its function—ceases to be so mysterious. As with threads into cloth and wires into networks, writing took lines and lives and made them into something different. It offered, in words repeated throughout the Account, "Order and Method" (6). These women didn't want to *feel* better; they wanted to *be* better.

Pleasure, aesthetic and otherwise, was a secondary issue—an effect of the methodizing power of writing. Thus the object that elicits the greatest pleasure in the *Account* is neither a text nor a painting nor a song but the rewritten bodies of the Club members themselves:

> You cannot imagine, Sir, the Joy we had when we found our selves conveened in the Character of Members of THE FAIR INTELLECTUAL-CLUB. For my part I thought my soul should have leapt out of my mouth, when I saw nine Ladies, like the nine *Muses*, so advantagiously posed. If ever I had a sensible Taste and Relish of true Pleasure in my Life, it was then. (11)

This is improvement: improvement through interfacing with a technology.

In Douglas Engelbart's terms, this was "augmentation"—improvement generated by what he calls the "matching processes" that allow the human and the technology to alter each other. But what is the basis of this particular matchup? These women clearly understood writing to be the tool that transformed them—that gave to them what they really wanted—"the Character of Members." But for us to understand how writing accomplished this feat, we need to be more specific, for writing can take other guises and do other kinds of work. What kind of writing are we talking about here? What genre informs and contains all of the other genres of this written account and of the activities it describes, from letter to verse to scheme to constitution to harangue?

The formal feature that gives away the genre of the Fair Intellectual-Club is the one I have already cited: the insistence, in the very first sentence of its Rules and Constitutions, on "Order and Method." These are keywords of system in the eighteenth century—not system as an idea but as a genre, as in Johnson's definition of "system" as the "reduc[tion]" of "many things" into a "regular" and "uni[ted]" "combination" and "order."

The method and order characteristic of the genre of system not only describe but configure all of the desirable behaviors articulated by the club. The *Account* itself is structured by them. The sketch of the club's history is a preface to the list of constitutional rules and thus follows the standard form of written system noted earlier in regard to Godwin's *Political Justice*: an enumeration of "principles" embedded in expository prose (see figure 7.2). Those rules in turn clarify the need for and nature of the transcribed harangues that make up the rest of the *Account*: the inaugural speech by the first speaker of the club and an admissions address given by a new member. The former, in its own words, is about the need for "Order and Regularity" (13); the latter is framed by the secretary as an example of "Method" (25).

If we read them ahistorically—that is, without these generic features in mind—neither the speeches nor their frames seem particularly conducive to intellectual or sisterly harmony. In fact, they come close to confounding them. Far from encouraging reading and writing about the sister arts, "Mrs. *Speaker*,"[12] as she's called in the *Account*, spends most of her time warning against the dangers and hazards of such activity—the threat of *dis*order that they pose. The secretary's written response after the

6 *An Account of the*

and Expediency of what we were to do. The more Time we ſpent in thinking and conferring together upon the Meaſures we had laid down, we were the more cheerfully diſpoſed to adhere to them; in ſo much that, when the Time of Meeting came, we were all ready to accompliſh our Deſign, with the greateſt Hope of Succeſs and Expreſſions of mutual Love and Friendſhip.

THE Original of our CLUB being thus far fairly repreſented to you, Sir, it will not be thought unnatural in the next Place, to deliver the Contents of the Paper which we ſubſcribed, upon the firſt *Thurſday* of *June*, and which I here tranſcribe from the Original it ſelf; as follows.

The RULES *and* CONSTITUTIONS *of the* Fair INTELLECTUAL-CLUB *in* Edinburgh.

WE, whoſe Names are underwritten, being ſenſible of the Diſadvantages that our Sex in General, and we in particular labour under, for want of an eſtabliſhed Order and Method in our Converſation: And being ambitious to imitate the laudable Example of ſome of our Brethren, that make the greateſt Figure in the learn'd and polite World, in ſo far as we are capable and may reaſonably be allowed, by entring into a mutual Compact and Agreement, to act for the Intereſt and Improvement of one another, in our Meetings; have reſolved to eſtabliſh a Club, called,

Fair Intellectual-Club. 7

called, *The* FAIR-INTELLECTUAL-CLUB; and hereby declare our Aſſent, and Purpoſe to obſerve, (whilſt we are alive and unmarried) *The* RULES *and* CONSTITUTIONS, which follow;

I. THAT we ſhall maintain a ſincere and conſtant mutual Friendſhip, while we live; and never directly nor indirectly reveal or make known, without Conſent of the whole CLUB asked and given, the Names of the Members, or Nature of the CLUB.

II. THAT none ſhall be invited or admitted into the CLUB before her Name be propoſed in it, and her Merits impartially conſidered, and Allowance given by all the Members to have her introduced.

III. THAT none ſhall be declared a Member of our CLUB, before ſhe hath, in our Preſence, ſubſcribed her Name to the RULES and CONSTITUTIONS thereof.

IV. THAT we ſhall never admit more than Nine into our CLUB, whereof Five ſhall be counted a *Quorum* ſufficient to act in Abſence of the reſt, as if the Number was compleat.

V. THAT none ſhall be invited or admitted into our CLUB before ſhe be fifteen Years of Age, nor after her twentieth Year is expired.

VI. THAT altho' different Principles and Politicks ſhall be no Hindrance to the Admiſſion of Members into our CLUB, being

Figure 7.2
The rules and constitution of the Fair Intellectual-Club.

transcription is, to say the least, surprisingly tart: "I leave you to judge, whether or not the Author of it deserv'd the Chair." She then reassures the reader that "we have been to this Day as careful to maintain good *Order* in our Meetings, as she has appeared zealous in recommending it" (24–25, emphasis mine).

The new member's harangue is presented in a much more positive light, but it too has a surprising twist. Asking from the start why she has been chosen, the initiate dismisses any explanation that has to do with content—her particular virtues or their opinions of them—and focuses instead on the formal structure of the "Occasion": "Your Goodness and Charity have *put you on a Method* to try my Respect and Gratitude" (emphasis mine). In other words, now she owes them: "with the Aspect of receiving a Favour you have oblig'd me to be for ever grateful and obedient" (28).

When I found *An Account of the Fair Intellectual-Club*, I expected a tale of aesthetic pleasures, intellectual freedom, sisterly goodwill, and, quite frankly, a good time, but what I found was order and method—an insistence on "Regulation" (5) that curbs immediate and easy pleasures, demarcates proper subjects, and invites criticism. These could be attributed to individual psychology—perhaps an "innate character"—or, zooming out, to the disciplinary nature of Western institutions, but both of those explanations would miss a crucial historical point. New technologies transform social relations; how they do so depends on the forms the technology assumes. In the eighteenth century in Britain, writing, as I have been insisting, was the new technology and system was one of its hierarchically dominant forms. Eighteenth-century clubs, I am arguing, were the social incarnations of the genre of system, sharing with that genre the informing features of method and order.

The strangeness of that pairing evaporates when we remember Kevin Kelly's description of system as "anything that talks to itself," for that is precisely what a club does; you join it because you want to be part of something that talks to itself. Just as the self-regulating power of a thermostat systematically maintains the physical environment, so systematic interfacing in clubs maintains the social; temperature is stabilized in the former, while conversation is sustained in the latter. Clubs proliferate in the eighteenth century because writing brings, in the words of the *Account*, "Order and Method" into "Conversation." Written rules and constitutions, for example, allow clubs to be self-regulating in matters ranging from behavior to size, as in keeping the membership at nine.

The *Account*, then, provides all of the components necessary for Engelbart's "augmentation." As the women conformed themselves into the systematic behavior of a club, writing assumed the form of system that Newton's choice had helped to proliferate. Just a few decades after the *Principia*, the human and the technology meet (they match up) *in* system. From that matchup emerges what Engelbart calls the "energy flow" into the "outside world." That flow took these women from their secret interfacing into the "outside" in terms of both being ready for marriage and of being exposed in the public. At first that exposure entailed the publication of their writing in the form of the *Account* and then in the form of their poetry—and, as we shall see in the next section, into a different kind of club. But, per Engelbart's notion of continuous improvement, the "energy" they generated did not exhaust itself back then. After running underground during much of the nineteenth and twentieth centuries, the flow has now reemerged, and not only in this book. In 2014, the *Account* took the stage in the new form of a play presented at the Edinburgh Festival.[13]

SYSTEM INCARNATE—DANCING INTO MODERNITY

Clubs, then, preceded culture as one of modernity's incarnations of system, as did a number of other social forms that enact that genre's internal conversation of parts as a whole. In fact, Johnson defined conversation at midcentury by distinguishing it from conferencing. Conversation, he asserts, is "familiar discourse; chat; easy talk: opposed to a formal conference." "Conferences" were not exclusively formal matters until roughly that point. The word had sometimes been applied to arranged meetings, such as the Hampton Court and Savoy conferences, but it more commonly referred to the act of conferring in the more general sense of conversation or talk. The restriction to formal discourse, codified by Johnson—"the act of conversing on serious subjects, formal discourse"—coincides roughly with John Wesley's calling of the first Methodist conferences in 1744.

The conference is such a common and sedate form now—a bulwark of our modern bureaucratic societies—that it is hard to recover the apprehension it was once capable of evoking. Such gatherings, especially when infused with religious enthusiasm, raised the specter in the eighteenth century of disruptive combination. What made conferences work, and eventually domesticated them, was that they were a solution to a specific historical problem: how to put the oral into a mutually productive relationship to the increasingly dominant technologies of print. "Conversation" became the general eighteenth-century rubric for this enterprise, and the conference was a specific version of that effort. In regard to religion, for example, Protestantism arose in relationship to print, its authorizing image the individual reader. Wesley enhanced that authority—and formalized it—by turning those readers into conferees. Methodism, then, can be seen as what conversation did to Protestantism.

All of these efforts were fueled by the proliferation of system, for method, as the *Account* hammers home, was a primary eighteenth-century keyword of system. To methodize was to arrange in a system—Methodists, for example, altered the religious landscape in the eighteenth century by instituting formal procedures for talking to themselves. Conferencing led to consensus—private individual readers coming together to speak publicly with one voice—a commodity of central importance to that century's formation of new forms of nationalism and capitalism, and thus an outcome that eventually took the radical edge off conferencing.

This phenomenon recurred throughout British society. The arranged demonstrations and formal correspondence of the Royal Society during the eighteenth century, for example, served the same function as the conferences of the faithful; to meet the society's new epistemological demand—"on the word of no one"—they methodized conversation to generate consensus: on the word of all. "Objectivity" is but our modern gloss on that consensus.

Improvement in the eighteenth century—whether for the Fair Intellectuals, Wesley's conferees, or the witnesses and correspondents of the Royal Society—was what Engelbart would call a "system-engineering problem" (1962, 5). The stakes for that kind of problem, he argued, were high: "the whole interface thing can change the very language and the very structure and the very modes [in which] we portray our symbols and communicate and think."[14] The club was the Fair Intellectuals' "thing" because it was where the need of the technology to propagate itself met the need of these women to change. They turned to it with the "thought" that women "who excell a great many others in *Birth* and *Fortune*, should also be more eminent in Virtue and good Sense" (3). One of the club's social functions was thus to provide the privileged with new markers of privilege at the moment that the old ones were losing their efficacy. As established forms of identity eroded, these women took on a new character—the "Character of Members." With that transformation, their souls, as the *Account* tells us, almost leapt out of their mouths. Their ecstasy was not pharmaceutical, but it was generic and technological—and warnings against the dangers of dependence on the then new technologies of writing and reading (especially by women) were as commonplace back then as antidrug messages are today (Siskin 1998a, 175–187).

Today's new electronic technologies have also assumed the form of systems—operating systems, network systems, telephone systems, ecosystems—facilitating, in turn, the formation of new kinds of clubs—from listservs and chat rooms to Facebook and LinkedIn. There, too, issues of gender and privilege, secrecy and privacy, mix with new kinds of order and method. In *Zeros and Ones* (1997), Plant pushes the possible parallels in her book in sometimes surprising ways, finding prototypes for change in both the new forms of nocturnal clubbing that proliferate in cities across the globe and in current reshapings of knowledge—and then compares the two. "Multidisciplinary research," Plant writes,

> like multimedia, is only the beginning of a process which engineers the end of both the disciplines and the mediations with which modernity has kept exploratory experiment under wraps. (198–199)

Many scholars, of course, have speculated that the standard disciplines are on the way out—but not out *onto* the dance floor. Yet that is precisely where you will have to join me as we follow the fate of the Fair Intellectuals. To my astonishment I discovered that it was not just the souls of the Fair that "leapt" about in early eighteenth-century Scotland but their feet as well.

In a 1724 issue of the *Plain Dealer*, a London periodical, Aaron Hill, the editor, and Fergus Bruce, a correspondent from Edinburgh, discuss the new forms of literary activity and conversation in the North. Their focus is on the phenomenon of clubbing and its connection to improvement and to gender. "Not the Gentlemen alone," observes Hill, "but the very Ladies, of Edinburgh, form themselves into select, and voluntary, societies, for the improvement of their knowledge."[15] After quoting from the *Account of the Fair Intellectual-Club*, Hill promises to

> say more, on a future occasion, of the honour done to the whole sex, by the dangerous Ambition of these ladies—And of the political Necessity, which, I conceive, there will soon be, of putting a stop to the progress of such unlimitted improvement of a power, already too exorbitant.

That occasion for further comment never arrives, however, for in the next month's *Plain Dealer* a new letter from Bruce diverts the discussion to the other club making waves in Edinburgh. Five years after the Fair Intellectuals wrote themselves down, setting "a pattern to Female Excellence," the "Fair Assembly" stood up—and *danced*. This new club, writes Bruce,

> consists of our best—bred ladies, of different qualities & ages ... all our Pulpits rail'd against it. ... But the Holy Fire is now much spent, & we are at liberty to

> meet in our great hall, without danger of the Kirk's Anathema; nay, some of the wives and daughters of the sanctified, begin of late, to grace our Fellowship. For my own part, I despair not to see the Reverends themselves, eating sweetmeats in our company: And mixing innocently, in our country dances. (*Plain Dealer*, 1724, 1.55.28 Sept.)

If, Bruce speculates, the social mix expanded even further to include the "sanctified" husbands and fathers themselves, "slander & detraction" would cease.

The full fantasy that follows is rather shocking—a sort of footloose Kirk—and yet it fits precisely the aspirations to mutual improvement of early eighteenth-century Scotland: "notwithstanding they are worthy Ladies, of undisputed virtue & honour, who preside over the Fair Assembly, I should be better pleased to see at our head, a moderator from the General One"—the General Assembly of the Church, that is. With the "Fellowship" of the Fair under the same watchful—but now sympathetic—eye of those who governed the Church of Scotland, improvement would know no bounds: "Husbands would allow their wives to go into company, without Jealousy, and Parents send their daughters, without fear of their leaving behind them any thing, that they *ought* to bring back again." This vision of pleasurable innocence is too much for Bruce, however, and he retreats to a claim for more modest gains: "But till that Halcyon day arrives, we must be contented with the want of sanction, and dance, and drink tea without them."

As with the Fair Intellectuals, clubbing is transformative: those who join in take on the "character of Members," and Bruce has a very particular "character" in mind. The Fair Assembly, he enthuses, is "one of the best nurseries of politeness in Scotland." The newly reborn

> learn a habit of briskness and freedom, which, till of late, were less frequent among us. That restraint, and stiffness of air and Address, which distinguished us from People of more generous education, by degrees, wears away; and conversation which was formerly so confin'd, and so dull, grows open, free, and easie. (*Plain Dealer*, 1724, 2.65.2 Nov.)

As we know from American Express, things are easier when you are a member—especially when joining one club gives you entry into another. The members of these early eighteenth-century clubs in Scotland wrote and danced their way into another form of union—one that had been officially formed in 1707, seventeen years earlier. "In short," concludes Bruce, "we exchange the Spanish Affectation and Gravity, for the English Liberty and Free Spirit … [this] conversion … is, in a great measure, the effect of the union of the two Kingdoms." Clubbing at the national level can be exquisitely ironic: Scots helped to invent Britishness by constructing their own version of Englishness (Crawford 1992).

Forty-three years later, the *Encyclopaedia Britannica*, the first encyclopedia to use a national allusion, emerged from Scotland as a sort of user's manual for the nation as a scaled-up improvement club. Like the Fair Intellectual-Club before them, Scotsmen such as Adam Smith performed the work of writing in an ongoing drive for method and order. To close the circle of the Union—to have Scots and Scotland take on the character of members—they too turned to system. That genre became the interface of choice between an enabling technology and social change. Just as those young women transformed themselves into a club that was a social incarnation of system, so Smith joined their dance and became, as we have seen, the Master System-atizer.

Douglas Engelbart's own interface "thing" is now, in one sense, everyone's: he methodized the use of computers by making us part of a system that mastered all operating systems—the windows and mouse desktop. But that was not the *real* thing for Engelbart. His central goal in his moment of technological change in the late twentieth century was the Fair Intellectuals' purpose in theirs: augmentation—in his case, the "augmentation" of that figure in the diagram: what he called the "human system."

The most famous interface tale of our times is the appropriation, first by Xerox and by Steve Jobs of Apple, and then by IBM, of Engelbart's means but not his end. Taking as their target users not human systems but office workers—defined as those interested more in tasks than technology—those companies shifted the interface agenda to "invisibility." Jobs famously argued that a computer should as easy to use as a toaster. Under the mantra of "ease of use," the technology was supposed to disappear. In discussions of the desktop, augmentation gave way to accommodation.

Engelbart's mission, which began as a secret—a secret report to the Stanford Research Institute under contract to the US Air Force—became a secret again once the mouse was domesticated. Until his death in 2013,

monday afternoon

december 9

3:45 p.m. / arena

Chairman:

DR. D. C. ENGELBART

Stanford Research Institute
Menlo Park, California

a research center for augmenting human intellect

This session is entirely devoted to a presentation by Dr. Engelbart on a computer-based, interactive, multiconsole display system which is being developed at Stanford Research Institute under the sponsorship of ARPA, NASA and RADC. The system is being used as an experimental laboratory for investigating principles by which interactive computer aids can augment intellectual capability. The techniques which are being described will, themselves, be used to augment the presentation.
The session will use an on-line, closed circuit television hook-up to the SRI computing system in Menlo Park.
Following the presentation remote terminals to the system, in operation, may be viewed during the remainder of the conference in a special room set aside for that purpose.

Figure 7.3
Original announcement of the public debut of the computer mouse, 1968.

he was intermittently celebrated but his agenda largely ignored. When his name did publicly surface, it was not to engage his notion of augmentation but to puzzle over why it took over two decades for his work to be realized (from 1962 to the debut of the Macintosh computer). When he surfaced surprisingly in *Sync* magazine in 2005, for example, a gadget magazine primarily for adolescent males (launched in 2004 and gone in 2005), it was in the form of a photo of him looking somewhat forlorn at eighty years old with the caption "He Didn't Make a Dime," a mocking reference[16] to his failure to cash in on the graphical desktop he invented.

The trajectory of his tale has thus strangely followed that of the Fair Intellectuals. Their club began as a secret, was leaked to an opportunistic male, and then became a secret again, made invisible by the very technology it championed. Similarly, the copy of the *Account* I found disappeared into an anachronism of print culture, the largely unindexed depths of a bound miscellany,[17] and writing itself veered during the long eighteenth century from being experienced as a prescriptive and thus enabling and potentially dangerous technology to a standard feature of culture—the "safe" tool of Literature (Siskin 1998a, 7, 201–203).

To bring the club and Engelbart back to light together is to reveal yet one more secret: the purchase their legacies still have on us. We are all still, in important ways, members—many of us as subjects of modern nation-states underwritten by print, and some of us as members and products of the disciplinary departments descended from the methodizing intellectual clubs of the eighteenth century. We are thus all still candidates for changes in character configured by system—whether and how those changes will make our hearts leap is another question. The reason for Engelbart's work taking so long to be realized also points to the ongoing possibility of change. The answer, quite simply, is that it still has not been realized. The "principles" demonstrated on December 9, 1968 (see figure 7.3), have altered the world and how we know it, but the role of system in ongoing change is far from over.

CODA

Re:ENLIGHTENMENT (ALGORITHMICALLY ENHANCED SYSTEMS)

"THE WHOLE ENCHILADA"

"Send your best and brightest," read an e-mail invitation sent by a MacArthur Award–winning neurobiologist to every department on my former campus, "to an informal, bi-weekly seminar whose purpose this semester will be the restructuring of the current organization of knowledge." This appeal stood out for its ambition and confidence—"this semester"—but the goal of restructuring knowledge has appeared on many agendas since the late twentieth century. The blame that has fueled the will to truth in the modern disciplines—driving their embedded systems to include more and more things so as not to fall short—appears to be assuming the form of desire: a call from inside and outside the academy for new forms and combinations of knowledge—for what is most often termed interdisciplinarity. From the proliferation of interdisciplinary humanities centers to E. O. Wilson's dream of "consilience"—a leaping together of the disparate disciplines (Wilson 1999)—to this invitation to take on the "Whole Enchilada," blaming has given way to hoping for a new resolution. As with the blame, the timing and intensity of this desire are keyed, I will argue, to how system works within our organization of knowledge. The change it signals is taking shape—like the previous shifts this book has described—as a redeployment of system.

My experience with the Whole Enchilada group provided the first clue as to the nature and timing of that change. In culinary terms, an enchilada is about adding in the chili. The secret to making a whole enchilada is to leave nothing out. The invitation to this one was an invitation to totalize—to put this more precisely, to totalize knowledge in a different way. As department chair, I sent myself to reserve a place for my discipline in this new whole. What neither I nor the organizer nor the other participants even entertained back then—at the turn into the twenty-first century—was that "discipline" itself, as the assumed shape of knowledge since the time of *Britannica*, might itself be a candidate for change. While interdisciplinarity preserves discipline as that which is being mixed or shared, the possibility of a reshaping of the basic unit of knowledge was not then on the table.

As with the recipe for a whole enchilada, we added everything into this group. The first few meetings, variously attended by historians, sociologists, psychologists, biologists, physicists, medical doctors, engineers, and others, were not promising. At best we enacted the interdisciplinary norm: wary encounters at a distance, punctuated by the borrowing of desirable content. But, then, what was only a whisper from across the table at the start of the semester became a chorus. Darwin was somehow the lingua franca of this group, but exactly how and why was not clear. The portability of "evolution" was certainly a significant factor, but since that porting had already been going on for quite a while, there had to be something else in play that more fully explained the connection between Darwin and our totalizing.

The explanation that follows was not the consensus of that group—and even now, since it entails a turn to the present and future, this account is necessarily more speculative than the ones I have offered for earlier historical moments. It benefits, of course, from what I have learned about the past in writing this book and what has happened since the group met. But what has been most crucial to mounting this explanation has been the positing of different kinds of histories. As an alternative to the history of ideas, the history of mediation tells us to look to form as well as to content, and the history of the real, although a necessarily blunt instrument, scales up our perspective on how forms change and how change functions over time.

My primary clue back then for identifying Darwin's purchase on the group was a book published just a year before we convened. The philosopher Daniel Dennett, in trying to explain why he considered Darwinian thought to be a "universal acid," altering all that it touches, came up with a catchy—but I think misleading—title: *Darwin's Dangerous Idea* (1995). The real catch in his argument is that he defines the idea in two different ways. At first, he claims that natural selection is the idea. But then he complicates matters by calling natural selection an "algorithm." Darwin, he points out, actually wrote up natural selection in the same "If … then" format that we use today in computer programming. The dangerous idea is then significantly reworded as follows: "All the fruits of evolution can be explained as an algorithmic process" (60). Without exception, in other words, every manifestation of life on earth is a product of the "mindless mechanicity" of the algorithm of natural selection running again and again; no "skyhooks"—Dennett's term for any form of external help, such as God, beauty, or manifest destiny—allowed (74).

In the first version, the danger lies entirely in Darwin's particular subject matter—what natural selection *means*. In the second, it shifts toward a more general and consequential issue: what an algorithm *can do*. That shift, I would argue, is crucial, for it is what makes the acid universal. The importance of natural selection for knowledge beyond the biological lies not in the content but in the form it assumes, as well as in what that form—as form—is capable of performing. If, in the instance of biological diversity, a simple algorithm can account for the greatest complexity, then it would be important to identify other instances of this formal phenomenon.

The clear and present "danger" in Darwin, I am claiming, lies not in natural selection itself but in that phenomenon. An algorithm, in Dennett's words, is a "certain sort of formal process" (50)—that is, a process written up in a certain form. Although the idea of evolution has been very powerful in general and certainly was for the Enchilada, Darwin's purchase on the group's drive to wholeness—the primary danger it posed and continues to pose to the "current organization of knowledge"—is better understood as a matter of a form and its effects, of mediation by a genre.

The genre at issue here may at first glance appear to be unrelated to our generic focus on system. However, as Dennett's use suggests, "algorithm" can and should be engaged as more than local to maths and computer science. In fact, even in those fields, algorithm and system are deployed more generally as variations of each other. In a semantic sense, for example, *The Free On-line Dictionary of Computing* defines *system* as "any method or algorithm" (foldoc.org, accessed Oct. 26, 2013). Seeing their variation from each other as a matter of degree rather than kind highlights system's scalability and thus its historical role in negotiating the relationship between complexity and simplicity. To put this in more practical terms, if we reverse-engineer any complex system, we will find—per Dennett's example—the simple algorithm that generated it.

Algorithm is thus a necessary part of our history of system as a genre, and we can tie it in by connecting Darwin to our earlier analysis of the role of system in Malthus. He claimed that the principle of population had evaded even "the most penetrating mind" because of things that had been left out: "the histories of mankind that we possess are histories only of the higher classes." With the lower classes included, the regular oscillations of population could be detected. Translated from history into system, their regularity became a "law of nature" (1798, 32, 15).

Sound familiar? Here is the more famous biological version: expand the histories of life on earth to include lower species and thus make visible previously undetected patterns that suggest a natural principle governing all forms of life over time. Like Malthus, Darwin raised the ratio of system by making the principle generically travel—in his case, he also literally traveled on the *Beagle*—to include more things. He did so by innovating on system, deploying it algorithmically as his new mode of inclusion. Whereas the standard account of Darwin's debt to Malthus features an idea—that Darwin was struck by the notion that population pressure would induce a contest for survival—the history of mediation calls our attention to form. The systematizing of political economy that depressed Malthus's contemporaries with the prospect of a starving world enabled Darwin.

We saw earlier how Malthus's *Essay* exemplified the power of embedded systems—of how that generic mix became newly masterful by avoiding the liabilities of the all-inclusive Master Systems that preceded it. Darwin's work, in turn, became an example for us, for it addressed the new set of liabilities that emerged from embeddedness. Unlike Malthus, he did not take those problems of fit and incompletion as an occasion for apology or revision; rather, he took it on as an opportunity to make his case in the manner Dennett describes, betting his system on the still unproven reach of its informing, algorithmic principle. "*If* it could be demonstrated," he wrote, "that any complex organ existed, which could not possibly have been formed by numerous, successive, slight modifications, my theory would absolutely break down" (Darwin 1909, 46, emphasis mine). By asserting not the truth of his theory but the conditions under which it could be proven wrong, Darwin lured even his opponents into the tasks of accumulating more examples and of trying to fit them to evolution. Together, they raised the ratio of the system by increasing the numerator—the number of things explained.

That placed the pressure, of course, on the denominator: would more principles be required? Newton let God haunt his system from the outside as a final cause. Malthus initially worked differently, making his system travel, but once it reached the crucial juncture—where population was supposed to explain morality—he opted for the same skyhook: God engineered the population problem, he claimed, to turn the earth into a hornbook for the soul, testing and teaching us through suffering (Malthus 1992, 79, 241, 385). Darwin, however, introduced a different mode of amplification: by habitually articulating his informing principle in the conditional "If … then" format, he harnessed the amplifying power of repetition. Every time the condition is met, the principle reasserts itself; thus the more it travels, the more powerfully it resonates.

Algorithm thus played a major role in Darwin—so major that we need to focus in on a twist in how it was deployed. Actually, it was more a doubling than a twist—a doubling that echoes the doubling of system I highlighted at the start of this book. We speak of system, I observed then, as something both conceptual—a way of knowing the world—and as something that is really there, that constitutes part of that world. In pinning down Dennett's sense of danger, we have enacted the same phenomenon in describing the algorithmic: we engage it as both a formal mode of producing knowledge about the world and—thanks to its simple-to-complex "mechanicity"—as constitutive of that world. In a sense, the world takes shape in the same way that knowledge does: both are products of an algorithmic system.

This is neither error nor ambiguity in Dennett's argument or my own. Doubling of this kind—think of our earlier example of Adam Smith reacting to Newton's system "as if" it were "real"—points to how the structural resemblance between knowledge and its object can signal that the object is indeed knowable. In the Enlightenment, it generated the conviction that the world could be known—known because the representations we call knowledge seemed to resemble it so closely. One way to grasp the importance of Darwin's turn to algorithm is to see it as putting real pressure on that "seem." In his algorithmic use of system, Smith's "as if" becomes a superfluous buffer. Instead of just an object to be known, the world (or nature in Darwin's case) becomes the *product* of a form of that knowledge: the simple instruction that we call an algorithm.

Strangely enough—or not so strangely, given the focus on "life" they shared with Darwin—that is what Watson and Crick's discovery of the structure of DNA confirmed a century later: genes carry a set of simple instructions, written in the repetitive syntax of the algorithmic, that are read off for purposes of reproduction. In David Deutsch's words following Richard Dawkins, genes "embody the knowledge necessary to render their organisms." "What the phenomenon of life is really about," he claims, "is knowledge" (Deutsch 1997, 180–181).

Deutsch uses the term *virtual* in a very specific way to up the ante in considering the alignment of knowledge and the world:

> Our external experience is never direct; nor do we even experience the signals in our nerves directly—we would not know what to make of the stream of electrical crackles they carry. What we experience directly is a virtual-reality rendering, conveniently generated for us by our unconscious minds from sensory data plus complex inborn and acquired theories (i.e., programs) about how to interpret them. (1997, 120–121)

Virtual reality refers here not just to computer simulations but to "any situation in which the user is given the experience of being in a specified environment" (122). The fact that it is possible, argues Deutsch, tells us something not only about our mode of knowing but also about the structure of that which is known. If "every last scrap of our knowledge" is encoded in programs for rendering the physical world on the brain's virtual reality generator, then we experience those scraps as knowledge because they allow us to survive in that world. "The fact that virtual reality is possible," observes Deutsch, "is an important fact about the fabric of reality" (122).[1]

Bear with me here, for this argument points to a new and important answer to the debate we have been tracking about the relationship of system to the world: are systems solely made by us in an effort to know the world, or is the world itself a system that we learn to discern? Adam Smith, as we have seen, opted for the former. Systems, he insisted, were "mere inventions of the imagination." The abbé de Condillac, however, declared just as forcefully that "systems should not be invented. We should discover those which the author of nature has made."

What bridged these two positions was the emphasis not on the nature of systems but on the nature of their makers: human or divine? What Deutsch's virtuality argument highlights, however, is the nature of nature itself. And that is where Darwin's algorithmic workings of system come into play. Should we take the doubling to its logical conclusion? Have we been able to know the world through system *because* the world is, in fact, itself structured as a system? And is it a system because system itself does the making? If so, then system would not be a product of man or god—as in Smith and Condillac—but would itself be a mode of (re) production.

Pushing the argument in that direction would pry open what we might call Darwin's door—the door to what the scientist Stephen Wolfram (2001) proclaims "a new kind of science." With Newton as its model, classical science and its relativistic variants have been using systems cast in the language of mathematics to try to approximate the real, constantly massaging the equations with new constants and particles to make them match increasingly stranger versions of that reality. For Wolfram and others—he has been roundly blamed for taking too much credit for this type of argument[2]—the running of system is not an approximation; it is the very nature of the real. The same, simple algorithmic systems with which we compute are not simply rules for taking the measure of the real; they actively account, he claims, for reality in all of its complexity.

I am neither advocating nor judging these related but often conflicting claims; Deutsch parts ways with Wolfram, for example, on what is "inherently simple" (Zenil 2013, 12694). I wish only to add them to my account of system's role in reshaping knowledge—and thus to clarify what may be at stake in making them now. While in the three primary parts of the book I recovered the earlier adventures of that genre, my job here is to report and synthesize its more current exploits. There are many of them, and they are hotly debated. In Wolfram's case, the complaints by fellow scientists and the media that he has hijacked the work of others to sell as his own have themselves hijacked the initial reception of his arguments. But once we recognize that both his claims and the acts of blame they have generated are episodes in much longer histories, then questions of a different kind—of form as well as of content—arise.

Why, for example, the big book? To rivals angered or intimidated by his 1,263 pages, Wolfram could have pointed out that it worked for

Newton—another scientist plagued by debates over originality, as in Leibniz's claim to the calculus. Newton, as I noted earlier, deliberately opted for huge, totalizing books over journal publication, a practice that was eventually reversed as the grand ambitions of Enlightenment gave way to essays, articles, and the other specialized practices of the disciplines.

Wolfram's scaling back up, however, should not surprise us, for it highlights how efforts at new kinds of knowledge require shifts in form as well as different ideas. "I gradually came to realize," recounts Wolfram, "that technical papers scattered across the journals of all sorts of fields could never successfully communicate the kind of major new intellectual structure that I seemed to be beginning to build" (2001, ix). The result is both an echo of earlier efforts to totalize and a departure from them: his big book has been heavily remediated by new technologies. Wolfram wrote and produced it in then new kinds of software—Framemaker using Mathematica—thus allowing for its extraordinary array and display of illustrations. But that was just the first step. The scaling up has been an ongoing project that exceeds the physical bounds of the book itself. *A New Kind of Science* has its own website, where the full text is free online, accompanied by complementary software as well as message centers, directories, and other forms of explication, communication, and publicity.

Why are algorithmic systems at the center of all of this attention *now*? System's presence at a scene of knowledge should certainly not surprise at this point in my argument. Neither should the timing if we focus on the machines that algorithmic systems animate: computers. Wolfram's entire project is based on the growing capacity of that technology to run algorithms over and over again, producing patterns of complexity from their simple rules. "Every detail of our universe," he asserts, "everything we see will ultimately emerge just from running this program" (545). He means "see" literally, for what emerges are patterns on the computer screen—patterns of cells that change color and number with each iteration of the algorithm.[3]

But—and this is the crucial point in understanding Wolfram-like arguments as a historical event—it is not the running of algorithms that is new; it is the *speed* at which the electronic and digital allow us to do it. Algorithms were, per Darwin, "run" by earlier technologies—and, despite the slower speed, to great effect. It made Darwin the start of something that led much later to both the convening of the Whole Enchilada and—just five years later—the publication of Wolfram's book.

I know how strange this connection sounds in terms of both its duration and its dramatic personae: 142 years between Darwin and Wolfram. But making that connection can help to illuminate how shifts in technologies entail shifts in form and thus in knowledge. Just as the proliferation of print in the late eighteenth and early nineteenth centuries contributed to the shift from Master Systems to the modern disciplines, so the dramatic increase in computer clock speed (the running of algorithms)[4] in the late twentieth century has occasioned claims to new kinds of knowledge.

If this connection is load bearing, then Darwin's legacy is clearly changing, extending far beyond the religious, philosophical, and scientific controversies of evolution itself. Its new reach can help us to identify the Enchilada's call for restructuring as yet another effect of the type of formal change I have tracked throughout this book—the redeploying of system. What does that type of change look like now? How would it feel?

One manifestation has been the new episode of clubbing I have described. As with the Fair Intellectuals, there was much pleasure around the table as we gathered together to take on the "character of members" of the Whole Enchilada. It was a pleasure linked both to the use of a relatively new and empowering technology—computation this time instead of writing—and to the sense that this change in character was also a resystematizing; per their nine muses, we were finding our places in a newly constituted whole. The shape of that whole was and still is unclear, and so keeping all options open requires an open-ended term for the desire that brought us together. Instead of *interdisciplinary*, with its baggage of preserving discipline, we might only need to set a vector for change. The term *dedisciplinary* points simply to a sense of moving from ("de-") the structure we have now, leaving open the possibility that even the basic building block of that structure—"discipline"—might itself change.[5]

The history of blame provides one way to frame that possibility. One reason that blame has accompanied system, I have argued, has been that system was understood to be something made—made to approximate the real and thus always falling short of it. Thus Locke complained, as we saw earlier, that

> though the world be full of systems . . . , yet I cannot say, I know any one which can be taught a young man as a science, wherein he may be sure to find truth and certainty, which is what all sciences give an expectation of. (1693, 229–230)

But what if now, in a way Locke could not have anticipated, system came to be understood as truth—its algorithmic repetition yielding the very certainty that science seeks? If, to push Darwin's door wide open, system was understood to produce the world rather than approximate it, could the disciplines still function as disciplines without system to blame for falling short? What if the effect of system traveling into those specialized fields was not the attenuated authority I described earlier but the sense of confident consilience that motivated the Enchilada? And what if system's new reshaping of knowledge is turning the epistemological logic of disciplinarity on its head, reformulating the once startling conviction that the world could be known through specialized, complex systems into the notion that knowledge—in the form of simple, iterative systems—renders the world.

My purpose here, of course, is neither to predict nor root for or against this future or any of the many variations of it.[6] Instead, I am extrapolating this particular one to bookend my turn back to Darwin with a look forward to what might be on the other side of the door—to the shape that Dennett's sense of danger and the Enchilada's ambition may yet assume. Doing so highlights the issue of duration—the apparent delay, that is, between what Darwin did and how we are realizing it now. For students of technology, such a delay would appear to be yet another example of new tools and technologies taking time to take hold—to mediate, as William Warner and I have put it, in the particular ways that we have come to expect. They do so only after assuming forms and functions that were not obvious, or even possible, when introduced. There is, for instance, Lisa Gitelman's often hilarious study of how telegraphy morphed from a visual and writerly tool that used electricity to record messages on paper tape to an aural and speechlike medium for electronic communication. The computer's long metamorphosis from Charles Babbage's programmed gears of the 1830s to the ENIAC's missile-targeting tubes of the 1940s to the iPhone is another now familiar example (Siskin and Warner 2010, 10, 120–135).

That delay corresponds in length and dates with the Darwinian one—a surprise at first given that we do not think of Darwin as part of the history of the computer. However, once we recognize what they have in common—the centrality of algorithm and its iterative potential—then Darwin's legacy might be seen to rival Babbage's.[7] Equally illuminating as both a precedent and a context for Darwin's fate is the delay with which I began this book. Galileo's message from the stars, received and sent 250 years before Darwin, broke the seventy-year-old logjam preventing Copernicus's legacy from being realized. His discovery of a second lunar system, I argued, put abstract debates about system finally and firmly into the physical and the plural—and thus into the history of system(s) this book compiles. This type of delay—let us designate it a Copernican Delay[8]—was less about finding a purpose for a technology than finding a technology adequate to a purpose. For Galileo, that meant an "improved" spyglass. For our engagement with Darwin, it means an improved processor—one fast enough to generate the message that Darwin, like Copernicus, may belong squarely within the histories I have brought to bear on system.

THE PERSISTENCE OF SYSTEM

If Darwin does land there, his turn to the algorithmic may mark the initial unfolding of the next chapter in the history of the real. Processors and processing have not only improved; they have also now taken over center stage in debates over the nature of the real. The year 2013 began with an 856-page volume, from over fifty authors, titled *A Computable Universe: Understanding and Exploring Nature as Computation*. Most of the contributions argue for and explore what the editor, Hector Zenil, calls "the informational nature of reality"[9]

The disagreements are sharp and deep, but one common theme is that there has been, in Seth Lloyd's words, a "paradigm shift in how we think about the world at its most fundamental level." That shift posits a universe no longer made up of atoms or just energy, but of basic bits of quantum information (qubits). It thus unfolds computationally—it computes itself into existence—and, according to Lloyd, out of it. In his own monograph, he uses an estimate of the number of qubits that make up that universe to calculate exactly when they will all be processed, and thus when everything they constitute—which is everything—will cease (Lloyd 2006).

This is stimulating stuff with much to pursue, but this is not the place to do it. I have turned to it in this coda because—no matter how these explanations play out—they tell us something very important about system. It persists. If the turn to information and computation signposts a change—especially a paradigmatic change—it is not because everything has changed or changed in the same way. Many ideas have been altered or replaced, but one of the primary forms in which we think them—the genre that has played such a central role in shaping our knowledge—is still recognizably present. System saturates this volume as thoroughly as it did the Enlightenment—the word appears over 550 times in those 856 pages. How it is being used, and thus how it mediates knowledge, is changing, but it has been holding its central position in the history of mediation.

I wrote this book to explore that persistence. In doing so I hope I have shed light on how and why it has persisted and on the different ways it has shaped and continues to shape knowledge. Clarifying those continuities and discontinuities can also make system of use to others as a marker of change in any of the areas of knowledge it has informed. The computational universe provides a telling example with which to conclude. "If I have one new message to convey in my book," Seth Lloyd declared in an interview, "it's that the universe is a system where the very details and structures in it are created when quantum bits de-cohere."[10] The primary continuity here is that system remains, as it was for Newton, the genre of choice—a genre still adequate to the universe. The discontinuity from Newton to Lloyd and Wolfram lies in the details of structure.

Structure is also where the discontinuity lay between Newton and his predecessors. The author of the *Principia*, according to Edward Grant in a passage I cited earlier,

> regarded his treatise as if it were revealing the mathematical *structure* of physical nature, rather than as the mere application of mathematics to nature, as virtually all previous natural philosophers would have perceived it. (Grant 2007, 313, emphasis mine)

A structural change of that severity had, not surprisingly, a domino effect on system itself. After Newton, Stephen Wolfram observes, "systems could only meaningfully be described by the mathematical equations they satisfy" (Wolfram 2001, 860). That is why, in a nutshell, Newton rejected his own first draft of Book 3 of the *Principia*, opting instead for a more mathematical account of the "System of the World"—a system described in terms of an equation for measuring the force of gravity.

To arrive at that equation, Newton set in motion the interplay of the calculus and system—of the dividing of wholes into parts and the assembling of parts into wholes. The result was a System of the World that proliferated systems in the world, giving shape, as I have described, to new kinds of knowledge. By changing the structural details of their System—information overwriting mass and energy per Lloyd and the foregrounding of algorithmic iteration in Wolfram—the architects of the computational universe are not only reshaping their knowledge systems but are also repositioning them. For "fundamental reasons," Wolfram claims, "systems, rather than traditional mathematics, are needed to model and understand complexity in nature" (860).

System and math are still at play here, but whereas Newton foregrounded mathematical principles—those are the first two words of his title, while *system* trails in Book 3—this restructuring puts system forward as the primary mode of understanding. The twist is that this is not really a turn from mathematics but a transfer of function; systems, in a sense, now do the math. Engaging system as a genre can clarify what has happened. As a historical grouping of features, the genre of system changes by changing the mix, and, starting with Darwin, the algorithmic has gradually become a featured feature of system. System is now coming to count in a new way.

Repositioning and repurposing system reconfigures, in turn, what I have been calling our knowable spaces. When Newton brought system into his treatise, choosing it as a primary form of guesswork, its subsequent popularity in the form of Newtonianism transformed the knowable spaces of moral philosophy, scaling them up into the Master Systems of the "human sciences." And when systems changed position again, traveling into other forms such as the essay, knowledge was again reshaped, yielding the modern disciplines. If system is now securing a new position in the computational universe—one in which it generates the world it helps us to know—then we may be entering a new chapter in the shaping of knowledge with system in a newly performative role.

APPENDIX A: NOTES ON VISUALIZATION

Mark Algee-Hewitt in consultation with Clifford Siskin

THE SCATTER PLOTS

These graphs are composed of points indicating the percentage of texts per year held by the *Eighteenth Century Collections Online* (ECCO) archive that contain (1) the word *system* anywhere in the text, (2) the word *system* on the title page, or (3) the word *essay* as compared to *system* on the title page. As the number of texts held by the archive varies per year, the use of the percentage (the raw number of texts per year that contain the term in question divided by the total raw number of texts per year) allows each year to be compared to the others. The overall presence of *system* or *essay* throughout the century can therefore be assessed. These numbers can also be read as measures of probability; for example, there is a 40 percent chance that a text drawn at random from ECCO's holdings for the year 1791 would contain the word *system* at least once.

To aid in interpreting the trends that each graph charts, each also contains a regression line, a prediction model that attempts to assess the pattern (or trend) of the data (whether it increases, decreases, or remains stable) and predict what the future of that trend will be. The equation shown on the body of the graph describes the characteristics of this line. In the graph "Percent of Texts Containing *System*," this is a linear equation (showing that the increase in the use of the word *system* was linearly progressive). In the other two graphs of title occurrences, the line is a higher-order polynomial, demonstrating the complexity of these data.

The graph titled "Percent of Titles Containing *System*" indicates an exponential increase in the use of *system* toward the end of the century: its data could also be fit to an exponential curve with little reduction in the goodness of fit. Below the equation, there is an *R*-squared number, giving the goodness of fit for each of the lines: this is a number between 0 and 1 that describes how well the regression model fits the observed data. An *R*-squared of 1 would indicate a perfect fit. These not only suggest the predictive power of the model, but also how regular the data are. With an *R*-squared of over 0.95, the linear graph is an excellent fit showing a high degree of regularity in the increase of the word *system* across the century. The graph of *system* in titles has an *R*-squared of 0.68, suggesting a good overall fit and a high correspondence to an exponential pattern of growth. Finally, the graph of *essay* in titles is the most irregular: the *R*-squared of 0.174 indicates that any observable historical pattern that may be present is remarkably weak. Unlike the graphs of *system*, this weakness of fit for *essay* indicates the nonsignificance of any increase or decrease in its use in titles throughout the century.

THE TECTONIC DIAGRAMS

The visual layouts in chapter 2 of the most significant words in the titles of eighteenth-century systems codes the probability of the co-occurrence of these terms as literal distances. We determined a word's significance by measuring the ratio between the number of times each word appeared in a title with *system* during a specific historical period and the number of times that we would expect it to appear during that period given its overall frequency in the corpus. We included only words that are relatively frequent in titles with *system*, but significantly less frequent in the overall corpus in the tectonic maps: this process allowed us to exclude very rare words that appear on only one or two title pages, as well as very frequent words, such as articles, pronouns, and modifiers, that appear on almost every title page. Both have little meaning in determining the unique clusters of words that surround *system*.

The visualization itself was developed through a new methodology based in part on work in information visualization at Bell Laboratories (Gansner et al. 2009, 345–346). Each map is based on titles that contain the word *system* within the ECCO archive. The term *system* therefore acts as the center for each diagram: all other words are placed in relation to both it and each other. Although in some cases it may appear as though *system* lies outside the visual center of the map, each map should be read with the tile that contains *system* in the center. In these cases, the borders of the map actually extend farther in the direction against which *system* is offset.

The distances between terms are visual representations based on the underlying relationships between the terms and are a function of the probability of their co-occurrence within a single title. The farther apart two words are, the less likely they are to both appear within a single title that also contains the word *system*. They closer together they are, the more likely they are to appear on a title page that also contains the term *system*. The cardinality of each term is determined by the relationship of that term and the word *system*. Groups of words that occur to the "south" of *system* share a different set of similar frequencies with each other than do groups of words that occur to the "north" of *system* on a given map. The specific location of each word on the map is a function of both how often it is likely to appear with *system* and what words it is most likely share a title page with. Words in the northeast corner therefore share a similar set of frequencies with each other that is different from those shared by words in the southwest corner. Each set of words may be equidistant from *system* (suggesting that both sets share title pages with *system* with the same frequency), but as the layout of the map is based on the relationship of every word to every other word, those in the northeast are much more likely to appear next to each other than to any word from the southwest corner.

In the early map, for example, the words *sir*, *isaac*, and *newton* all appear next to each other on the extreme eastern edge; this means that they are all equally likely to appear on a title page with *system* and much more likely to appear with each other than with a word from the western edge, such as *grammar*, which is more likely to appear with *language* and "*exact*. The absolute cardinality is arbitrary ("north" and "south" have no absolute meaning); rather, the distance and configurations of groups of words as they radiate out from the term *system* can be read as meaningful as these represent the relationship of each word to every other word. The size of each tile is a measure of its cluster density. The smaller the tile, the more tightly clustered it is and the more related to the words that surround it. The larger the tile, the less relationship it has to the words that surround it. The smallest tiles on each map therefore represent the words that are the most densely clustered and therefore have the highest probability of co-occurrence on the same title page.

Each map shows the configuration of the most significant words that share a title page with *system* in a different part of the eighteenth century. In dividing the period up into the three parts, we are primarily concerned with meaningful temporal divisions rather than dividing our total sample of 224,759 texts into three equal groupings. As the number of texts held by the ECCO archive per year is not constant across the century (reflecting the increasing number of books printed each year in the eighteenth century), our three maps are based on the following sample sizes: 43,703 for the early period, 61,559 for the middle period, and 114,497 for the late period. In each case, our results were scaled by the number of words on the title page and number of texts per period. The maps therefore represent the probabilities of co-occurrence rather than the raw values. A word that is rarely used, but when it is, it is always used with *history* would be closer to *history* on the map than a word that is used very frequently, but appears on a title page with *history* only half of the time. This latter type of word would appear with *history* more times than the former, but the probability of the former appearing with *history* is much higher, and therefore it would be placed closer to *history* on the tectonic map. This allows us to compare the maps to each other, even though each is based on a different number of texts.

The underlying relationships that structure the map are derived from the distance metrics that establish the relationships between words based on their co-occurrence in titles that also contain the word *system* in the holdings of the ECCO database. In short, each word receives a distance score from every other word based on the likelihood or probability that both would appear within the same title. Next, these data are used to build a network graph by reconfiguring this distance matrix as a series of nodes and edges. In the network, each word is represented by one node, with each node serving as the locus for a set of edges that extend to its nearest neighbors. The number of edges is limited by the distance of each node to its surrounding nodes: those within tight clusters of terms have more edges than those that occupy their own space. A force-directed layout of the network is then derived using a GEM (graph embedder) analysis of the data (Frick, Ludwig, and Mehldau 1994). This algorithm uses the node and edge information to simulate a physical system of springs whose equilibrium of forces is used to arrange the nodes within a meaningful and readable layout. The number of neighbors each node has and their distance determine the location of that node in relation to all of the others. In every case, *system*, as the only term that is related to all other nodes, remains the center point of the graph. Finally, the points at the coordinates determined by the placement algorithm are used to derive tessellated polygons in a Voronoi diagram, creating a topological layout of the network (Okabe et al. 2000, 43–45). The resulting map provides a more complex visualization of the relationships among the individual members and their local neighborhoods, replacing the raw distances with a depiction of relationality based on proximity, area and conjunction.

TECHNICAL NOTES

i. Although the ECCO archive contains many duplicate titles (through the inclusion of foreign reprints and later editions), we have elected to retain these duplicated titles rather than filter out the unique texts. As duplication is an indicator of a text's significance (only the most popular texts are issued in multiple editions), their inclusion allows us to more finely trace the cultural significance of our terms.

ii. The Tectonic maps presented in chapter 2 are based on a composite of two force-directed layouts. The first uses the scaled distance between nodes to derive the shape of the network. The second recomputes the layout using the initial placement of the nodes in the first graph (to which the GEM algorithm is highly sensitive) instead of the distance: this smooths the layout and establishes readable distances between the points. (See Fruchterman and Reingold 1991.)

APPENDIX B: TITLES CONTAINING *ESSAY(S)* IN THE PLURAL VERSUS THE SINGULAR

As noted in the main text, there is an upsurge in the number of titles featuring *essay* in the plural in our ECCO sampling in the last two decades of the eighteenth century. This increase appears to be tied to an increase in the number of volumes of collected essays. This poses the issue of the relationship of the genre of essay to the genre of collected essays. To aid in future inquiries into the realm of essay, I have supplemented the comparisons of *system* and *essay* in the main text with figure B.1 comparing the singular and plural forms of *essay*.

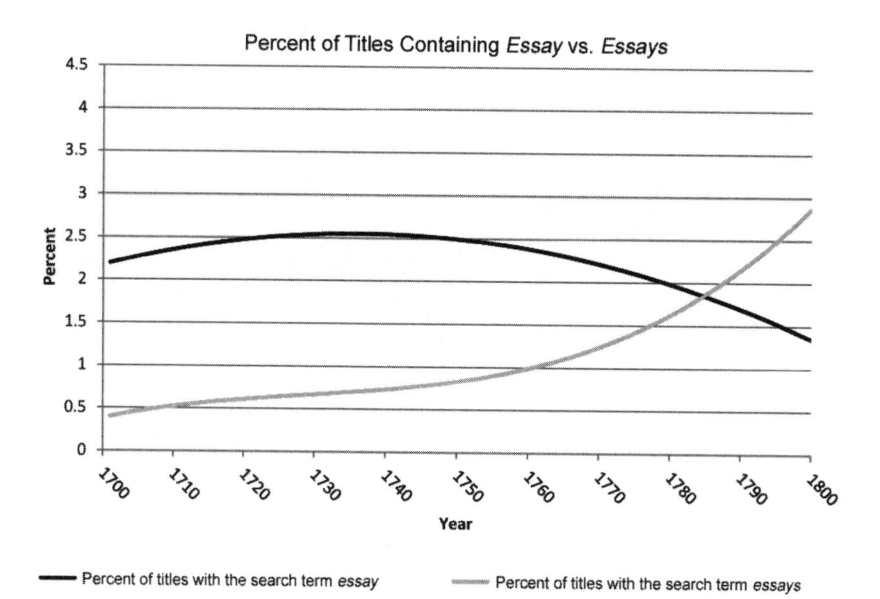

Figure B.1
Using *Eighteenth Century Collections Online*.

NOTES

PROLOGUE

1. Stanford University online lecture, January 9, 2012, http://www.youtube.com/watch?v=iJfw6lDlTuA.

2. An example of this distinction—idea/definition versus genre/explanation—in literary study is the trap of trying to provide an essentialistic definition of the "novel." Efforts to define it by asserting a set of fixed features—for example, it is a "long" form of fiction—are repeatedly undone by the historical nature of the things that have been called "novels" and the changing sets of features those things exhibit. In the case of length, early uses of the term *novel* often distinguished it from then longer forms, such as the ancient romance. Turning to kind, and thinking of the novel not as a fixed form but a genre, allows us to make these changes part of our explanation of what novels have been and are.

3. For the place of literary history in Bacon's scheme of knowledge, see Morrison 1977, 598.

4. The notion that there was, and is now, "more to say" is not a judgment on my part in favor of system over its historical competitors, such as hypothesis. It is descriptive of system's ongoing presence.

5. Galilei 2008, 68. I have used two different translations of Galileo (1989 and 2008) to best capture what I understand to be the meaning and force of his arguments. The 2008 work is a selected edition of Galileo featuring a recently revised translation of *Sidereus Nuncius* (based on the Edward Stafford Carlos translation of 1880) and a widely admired and convenient selection of works.

6. See Albert Van Helden's description of Galileo's publication strategy in Galilei 1989, 18. One consequence of this aggressive strategy was that Galileo's claims to originality were immediately challenged in terms of being preceded by the Pythagoreans and of having his "feathers" "plucked" by Kepler's *Dissertatio cum Nuncio sidereo* published in that same year, 1610. See Kaoukji and Jardine 2010, 441.

7. Bacon's *Of the Proficience and Advancement of Learning, Divine and Human* was published in 1605 and The Great Instauration appeared in 1620. For this quote, see Bacon 2000, 7.

8. See Van Helden's (1995) full explanation of the "delay" in fully engaging Copernicus in the Galileo Project: http://galileo.rice.edu/sci/theories/copernican_system.html

9. I take this use of *load bearing* from Peter de Bolla's description of "load bearing" concepts in Siskin and Warner, 2010, 89.

10. Robert Cummings has suggested the following translation: "Eximium praeterea praeclarumque argumentum habemus pro scrupulo ab illis demendo [We have moreover remarkably clear evidence for disposing of their scruples]."

11. This is a dedicatory poem to Michael Drayton by John Selden published in a book of Drayton's verse. Online OED accessed March 19, 2016: http://ezproxy.library.nyu.edu:2639/view/Entry/196665?redirectedFrom=system#eid.

12. For this dating of Enlightenment, see Siskin and Warner 2010, 18, 415–416.

CHAPTER 1

1. This diagram was constructed by Mark Algee-Hewitt at my request (see the Acknowledgments and appendix A for notes on these and other visualizations). Computable databases are the product of the late twentieth century, and the database used here, *Eighteenth-Century Collections Online*, came online in 2003.

2. McLuhan 1994. For a useful attempt to trace the genealogy of McLuhan's extension of the senses, see Rae 2009.

3. Simon Forman includes "glasses" in his list of new resources. See Forman 1591, image 6.

4. All heavenly bodies were referred to as stars. "Wandering stars" were planets. Galileo discovered stars that wandered around wandering stars—in other words, planets other than earth had moons.

5. See Maurice Finocchiaro's account of this exchange in Galilei 2008, 9–10. The full title of Foscarini's book is *Letter on the Opinion, Held by Pythagoreans and by Copernicus, of the Earth's Motion and S's Stability and of the New Pythagorean World System*. For a translation, see Blackwell 1991.

6. Ong 1956, 236.

7. See, for example, the constant references to system in Yates 1966 and Carruthers 1990.

8. The *Novum Organum* was published as the second part of the *Instauratio Magna* in 1620. Bacon specified that the "first part" is "lacking" but may "*to some extent*

be recovered from The Second Book of The Proficience and Advancement of Learning, Divine and Human" (Bacon [1620] 2000, 25). The various parts of Bacon's plan were republished in different combinations after his death. The *Advancement* of 1640, for example, is called on the frontispiece, "The Great Renewal Part 1" and includes the preface to the *Instauratio Magna*.

9. See Kaoukji and Jardine 2010, 237. They make the mistake, however, of referring to the 1620 frontispiece of the *Novum Organum* as the "engraved title page of the first, 1605, edition of Bacon's" *Advancement*.

10. For an examination of the use of *species* in the eighteenth century, see Ramos 2010.

11. Here is the *OED*'s full definition of this meaning of genre (accessed on June 8, 2012):

> b. *spec.* A particular style or category of works of art; esp. a type of literary work characterized by a particular form, style, or purpose.
>
> 1770 C. JENNER *Let.* 5 May in D. Garrick *Private Corr.* (1831) I. 384 With regard to the *genre*, I am of opinion that an English audience will not relish it so well as a more characteristic kind of comedy.
>
> 1790 A. YOUNG *Jrnl.* 15 Jan. in *Trav. France* (1792) I. 273 It is a *genre* little interesting, when the works of the great Italian artists are at hand.

12. See http://marshallmcluhanspeaks.com/understanding-me/1968-probe-as-a-tease.php and http://marshallmcluhanspeaks.com/understanding-me/1968-probe-as-a-tease.php, accessed June 10, 2012. Also see McLuhan and Carson 2001.

13. See my work on the quantitative issue of the late eighteenth-century proliferation of print in Siskin 2005, 797–823.

14. Per Walter Ong's description of title pages as once being "addresses to the reader" (Ong 1956, 229), most title pages through the eighteenth century included much more information than we find on them today. Much of that information consisted of attempts to attract readers by describing both the kinds and contents of the works. And since the range and properties of kinds were much less settled than now, those descriptions were often very full and featured overlapping claims about kinds. Such title pages are thus robust targets for electronic search today.

15. Boswell 1984, 176.

16. What is big for a big book has changed rapidly in the early twenty-first century. Many presses now press for no more than 90,000 words. Weinberger's strategy of translating the debate over the future of "the book" into a discussion of "lengths" of knowledge is quite useful in this regard. It gives us a context and vocabulary for efforts to innovate on the standard sizes. In 2012,

for example, Palgrave announced its "Pivot" initiative in the following terms: "Liberating scholarship from the straitjacket of traditional formats and business models, Palgrave Pivot offers authors the flexibility of publishing at lengths between the journal article and the conventional monograph." See http://www.palgrave.com/pivot/.

17. For a compelling example of using the shape of knowledge to help understand it, see Bazerman 1988. My focus on system here is complementary to his on the experimental article.

18. See the examples reprinted in Ong 2004, 200–202, 317.

19. In Bacon 2000, Michael Silverthorne translates *apparatus* as *organization* (20). In Bacon 1994, Peter Urbach and John Gibson translate it as *system* (24).

20. See also Phillipson 2010, 1.

21. For an examination of the notion of "completeness" in the eighteenth century, see Rudy 2014.

22. Walpole 1971, 604. Also cited in Wall 2006, 166–167. See Laura Yoder's 2012 discussion of collections.

CHAPTER 2

1. Focusing on system can help us better understand the consensus recently expressed in the *Stanford Encyclopedia of Philosophy* (2013) regarding Descartes: "for all of the forward-looking, seemingly modern, aspects of Descartes' physics, many of Descartes' physical hypotheses bear a close kinship with the Aristotelian-influenced science of late-Medieval and Renaissance Scholasticism," accessed January 2, 2016, from http://plato.stanford.edu/entries/descartes-physics/. For Bacon, what linked Descartes to the old organon was his reliance on the individual mind; for Newton, as we will see, it was Descartes' wrong choice of guesswork—he feigned hypotheses.

2. The earth's tectonic plates consist of oceanic and/or continental crust and the upper solid layer of the mantle—the lithosphere that "floats" on the asthenosphere.

3. I think of our relationship to this technology in terms of what I call the "paradox of fit": the better our tools perform the task we put to them and the questions we propose, the less likely it is that they will lead us to anything more than what we already know—they do not lead to new knowledge. That is not, course, a "bad" outcome; much of what we call "digital humanities" falls in this category. Figure 1.1 depicting the linear rise of system falls in this category. The point of such a tool is to make it easier to understand what we know—to simplify and settle issues such as whether system was used more frequently during the eighteenth century. The diagram proves that and presents that fact in easily

comprehended fashion. It fits. Tectonics, however, is a tool that does not enact the paradox. The technology becomes an active agent—not simplifying but cajoling us into new questions as we try to come to terms with it—it more than fits.

4. See note 14 of chapter 1.

5. Even if the plate is on the edges of the map, its presence indicates that there is a statistically significant relationship to system. See appendix A.

6. See Donald Kelley's claim that there was a "historiographical drive toward 'system'" in the eighteenth century (1997, 230). Kelley has been one of the best historians of system after Ong, and his claim is both plausible and compelling. But the visible evidence for it is a small number of telling quotations, and it turns on a metaphor. What is *drive* in this context? Does it need to be figured as personification? I ask not out of skepticism, but out of a desire for more evidence of a different kind—evidence that can materialize this drive on a broader evidentiary base.

7. Another new wordplate appearing for the first time on this third map is *systematic**. The fact that it has surfaced as a separate plate indicates that there are more title pages with just those adjectival and adverbial forms of *system*. This would, of course, be what we would expect in the century's closing decades as systems and features of system were increasingly embedded in other forms.

8. The source for this *OED* quotation is Brown, Webster, and Sexton 1903, 178; *OED* online, accessed June 14, 2012.

9. See Siskin 1998, 79–99, and Siskin 2009, 115. See also Gossman 1990, 41–45; Court 1992, 13–59; and Michael 1987.

10. See https://sysbio.med.harvard.edu/, http://cgsb.as.nyu.edu/page/about, and http://en.wikipedia.org/wiki/Systems_biology, all accessed June 18, 2012. As a new and diverse field, not all of the endeavors labeled "systems biology" fit Woese's prescription, but work on systems is playing a key role in the turn he advocates toward the study of "emergent patterns of organization."

11. Consolidating periods as in using "early modern" for a sprawling Renaissance, and adding as in turning the twentieth century into, among other schemes, "modern" and "postmodern."

12. For an analysis of old and possibly new tools of literary history, see Siskin 2007, 126–129.

13. See the arguments about the computational universe in the Coda. See also Deutsch 1997a, 95–96. Most recently, Deutsch has pursued his work on the relationship between knowledge and the physical under the rubric of "Constructor Theory." See Deutsch 2012, 24.

14. For a detailed analysis of this phenomenon, see Siskin and Warner 2010, 12–19.

15. By no means, however, have I been working on my own. See the acknowledgments for the details of the collaborative efforts that helped to generate this book.

CHAPTER 3

1. http://www.calvin.edu/academic/phys/observatory/images/Astr110.Fall2005/DeFrain.html.

2. In *This Is Enlightenment*, Bill Warner and I argue for Enlightenment as an event that stretches from a "clustering sense of difference" detectable in Britain, France, and Germany in the early 1730s to a closing moment of "saturation" in the 1780s. See Siskin and Warner 2010, 16–21.

3. See the explanation of "method" as "rhetoric" in Bacon 1994, 96, note 81.

4. The invention of the calculus was the subject of Newton's long dispute with Leibniz.

5. Another way to think about this connection between system and the calculus is through the use of system in music. The "Ancients," according to Ephraim Chambers's entry on that topic, called "a simple Interval, *Diastem*, and a Compound one *System*." A system, he concludes, "is properly an Interval, which is actually divided in Practice. … The Nature of a *System* will be very plain, by conceiving it an Interval, whose Terms are in Practice, taken either in immediate Succession; or the Sound is made to rise and fall, from one to the other, by touching some intermediate Degrees" (Chambers 1728, II.166).

6. "Vera autem Systematis constitutio sic pleniùs et exactiùs intelligetur." Quoted from *De motu Corporum Liber Secundus* (MS Add. 3990, Cambridge University Library, Cambridge, UK), http://www.newtonproject.sussex.ac.uk/view/texts/normalized/NATP00305.

7. See Newton 1978, 232, and 2004, viii.

8. I. B. Cohen recounts this guessing game, but sees the rivalry as the whole story: "In 1686, Newton was so angered by Hooke's demanding credit for the inverse-square law that he told Halley he intended to suppress book 3." See Newton 1999, 21, 48. The letter to Halley is Newton [1686] 1960, 435–440.

9. The emotive language of letters (*deserting*, *unkind*) plus the era's proclivity for personification do invite a personal and personality-driven interpretation of that sequence. The history of ideas, with the emphasis I have described on human agency, often yields interpretations that pursue that invitation. I turn to the history of mediation to focus instead on issues of genre and knowledge. These passages also lend themselves to a focus on gender, as well as on genre. Particularly useful in that regard are the efforts of scholars such as Sandra Harding and Evelyn Fox Keller to highlight the social values that permeate the language, concepts,

and history of science. See, for example, Harding 1991 and Keller 1985. I see my exploration of the Fair Intellectual-Club in this book as complementing their attention to gender.

10. For a best estimate of the sequence of composition of these versions, see Newton 1978, 236:

> Calmed by Halley's letter of 29 June 1686, and seeking "how best to compose the present dispute" with Hooke, Newton may have started pretty rapidly thereafter on the recomposition of the astronomical book "in the mathematical way" and accomplished this, as well as the completion of Book II, in the autumn and winter following. It seems more natural to believe that matters happened thus, than that Newton first wrote *De Systemate Mundi*, then abandoned its more popular approach and rewrote it, then decided to suppress Book III altogether, and finally changed his mind once more to include it. Indeed, his expressed intention of preventing those incapable of mastering Book I from following Book III may well have been specially aimed at Hooke, of whose mathematical attainments Newton had a low opinion.

This first version was published after Newton's death in Latin and in English in 1728. See Newton 2004, xii.

11. There is a history of simplicity of which Newton's effort can be considered a part. Aristotle, for example, begins the *Physica* with a paragraph invoking and centered on *στοιχείων*, meaning element but often understood and translated as "simplest elements" as in the following:

> When the objects of an inquiry, in any department, have principles, conditions, or elements, it is through acquaintance with these that knowledge, that is to say scientific knowledge, is attained. For we do not think that we know a thing until we are acquainted with its primary conditions or first principles, and have carried our analysis as far as its *simplest elements*. Plainly therefore in the science of Nature, as in other branches of study, our first task will be to try to determine what relates to its principles. (Aristotle 1930, 1, emphasis mine)

In his chapter of this history, Newton is setting up "simplicity" as a "rule" that addresses the particular issues of demonstrating hypotheses, doing so in the language of mathematics, and claiming a law applicable to the system of the world.

12. For the translation and a discussion of true and false systems, see Knight 1968, 52–78.

13. Cohen explains this translation as follows: "Newton here uses the word 'deitas,' a nonclassical term which signifies the essential nature of the divinity or 'god-ness.' Although 'Godhead does fit, the term 'godhood' (which is more abstract) may more accurately convey the sense of Newton's 'deitas.'" That level of abstraction fits, I think, with Newton referring to God as a "part"—and my thinking of that part as an optional topic for inquiries into natural philosophy,

despite Newton's own faith and the many words he devoted to theological concerns. See Newton 1999, 940, note 3.

14. The *Dictionnaire de Trévoux* appeared in various editions between 1704 and 1771, influencing both Chambers and Diderot. See Furetière 1704 for the first edition.

15. As long as systems remain objects of blame—whatever form the blame takes—the message about them will continue to be mixed; the mixture itself, however, will vary. For an early nineteenth-century example from Germany, see Tilottama Rajan's analysis of Friedrich Schlegel's "insistence that we must both have and avoid a system" generated by his notion of "the complementarity of unity and difference." Rajan 2000, 142.

16. Burton did not fend well in his rivalry with Smellie. For the details, see King 2007, 67–68.

17. For the quotation, see Walpole 1849, III.801. It was first printed at Walpole's own press at Strawberry Hill in 1771 and reprinted in 1785 in a limited edition, again at Strawberry Hill.

18. See the discussion of Smith and simplicity in Olson 1993, 106–107.

19. For the concept of prescription, see Siskin 1998a.

20. In Sloane's will, the word is used in its early modern sense, synonymous with "collection of objects," and refers to the things rather than the building that holds them. Older "museums" are particular collections, the work of a particular collector (Myrone and Pelz 1999, 193). In the late seventeenth and early eighteenth centuries, an emerging meaning of the word privileged the housing—the space—the place of the collection. Paula Findlen observes that by the end of the seventeenth century, "the language of collecting had evolved to such an extent that it designated the projects of collectors … as the 'museums' that gave meaning to the rooms in which scholars hoarded and displayed knowledge" (Findlen 1994, 49). Here, still, the word *museum* is significantly the collection, even as the meaning begins to leak from "projects" to the galleries. With the British Museum, the word came to signify to the public, national space in which objects were held and encountered. And the collection became something housed in a museum rather than the museum itself. The British Museum thus established a meaning of "museum" separate from a single collector and even, potentially, from a specific group of objects.

CHAPTER 4

1. As our own norms for knowledge bend under the pressure of technological, economic, and social change, our sense that they are ending turns our attention to when and how they began—and thus to how they differed from what came

before. This is another aspect of what Bacon called the good "fortune" of living in a moment of changing resources.

2. My argument here about the social function of knowledge in the Enlightenment, and my argument about the poet William Wordsworth as a system maker in chapter 6, complicate the standard period binary of Enlightenment Reason versus Romantic Reaction. This opens up notions of "Romanticism," as occupying the end of the eighteenth century and the first third of the nineteenth, to arguments about the transition into disciplinarity as the modern organization of knowledge. During that transition, elements we have come to see as opposites, such as art and system, mediated each other into new roles. My recovery of this chapter in the history of mediation complements other challenges to the period binaries, such as John Tresch's *The Romantic Machine*. He focuses on Paris in the early 1800s to show how, for a brief historical interlude, "emotional and aesthetic experience were valued on a par with technical and rational mastery" (Tresch 2012, 1). I show how "machine" and "system" overlap repeatedly in the histories I relate, from Bacon to Pownall to Darwin and the computational universe.

3. The notion that the principle of probability underlies our thoughts and actions at the neural level has, for example, reanimated approaches to certain kinds of artificial intelligence. Inadequately modeled by earlier turns to logic and neural networks, the brain in this new model works by calculating probabilities—a hypothesis vindicated by a computer based on probability algorithms vanquishing humans in a televised showdown on the game show *Jeopardy* in 2011. As David Deutsch (2012b) has argued, however, this is not a breakthrough in the more ambitious quest for AGI (artificial general intelligence).

4. This is from a paper Skinner delivered to the British Association in 2002: "Adam Smith: Science and System."

5. Let me acknowledge again, as I did in chapter 2, that my primary focus is Britain. Using this national lens makes these particular uses of system visible and sets the stage for future efforts to discover how system may have worked in similar and dissimilar ways elsewhere. The formation of the disciplines is one such marker. One way in which it will play a crucial role in this efforts is the manner in which disciplinarity, once instituted, counters the earlier geographical differences with a new kind institutional uniformity.

6. As noted earlier, the kind of "truth" is discipline specific. In the humanities, for example, Foucault described the will to truth as an "endless murmuring"—interpretation added to interpretation in an effort to complete the system. That effort, however, must be endless, because the price of success would be the end of the discipline. See Foucault 1998, 95.

7. For a descriptive analysis of the formation of the modern disciplines and of the historical connections between disciplinarity and professionalism, see Siskin 1998a.

8. The full title of the first, 1798 edition was *An Essay on the Principle of Population, as it Affects the Future Improvement of Society, with Remarks on the Speculations of Mr. Godwin, M. Condorcet, and Other Writers.*

9. See the examples provided in Wordsworth [1798] 1965, 319–326.

10. See, for example, the review of *Poems, in Two Volumes* published in the *Edinburgh Review* 11 (214–231), October 1807, reprinted in Haney [1904] 1970, 29.

11. Wordsworth (1798) 1965, 320. Southey's review appeared in *Critical Review*, vol. 24, October 1798.

12. Hays (1798) 2009, 8. I have removed the comma placed before "they involve" in this edition.

13. The early editions are undated. Glasgow's library dates the second edition as "1785?" Cambridge's library identifies "A new [10th] edition" as 1820. In the tenth edition, Mavor states that "more than thirty years are now elapsed since this System was first presented to the public."

14. Scolar Press has published a facsimile of all the texts involved as volume 18 (1980) of its Historic Accounting Literature series.

CHAPTER 5

1. Grady 1799, 53. He is commenting on the debate over the union with Ireland.

2. Only 100 of the 2,591 instances in *ECCO* occur before 1760.

3. For a discussion of these changes in relationship to the concepts of discipline and development, see Siskin 1988, 164–178.

4. The phrase "war of ideas" is Marilyn Butler's (1975). I use it here to emphasize how a history of mediation can complement a history of ideas.

5. These terms were overlapping alternatives for most of the eighteenth century. See Siskin 1998a, 172–174.

6. This sense of the loss of epic power in the present is, of course, a common theme in midcentury British poetry. In Collins's "Ode on the Poetical Character," for example, only Milton can wear the girdle of pure epic: the "inspiring bow'rs" have been "curtain'd close … from ev'ry future view" (1979, 34).

7. Barbauld is misquoting Andrew Fletcher in *An account of a conversation concerning a right regulation of governments, for the common good of mankind; in a letter to the Marquess of Montrose, the Earls of Rothes, Roxburgh, and Haddington. …* The actual quotation is as follows: "if a man were permitted to make all the Ballads, he need not care who should make the Laws of the Nation" (Fletcher 1704, 8).

8. For the historical career of the notion of the "rise of the novel"—of how that notion helped to constitute the novel itself—see Siskin 2015.

9. Since the *Review*'s primary purpose was to extol Scott as "the greatest master" (62)—the genre of the article is an encomium—it dates the change from *Waverly*.

10. From Godwin's Preface to the 1832 *Standard Novels* edition of *Fleetwood* (Godwin 1832). Excerpted in Godwin 1988, 350.

11. Four different letters were published in various combinations in different locations in 1784. I am quoting from a volume containing letter 1.

12. See Robert Mitchell's analysis of Smith's concern that the love of system not become "inhuman" (Mitchell 2006, 70). Mitchell also uses the love of system to link Smith to Samuel Coleridge (Mitchell 2005).

13. Polwhele's attack was titled "The Unsex'd Females" (1798). His point was to put them back in what he saw as their place: being females rather than writers. He wanted, that is, to re-sex them.

CHAPTER 6

1. Review of Wordsworth's Poems (1807) by Francis Jeffrey in the *Edinburgh Review*, cited in Woof 2001, 185–201.

2. Derrida and Ferraris 2001, 4. This first appeared in French and Italian in 1997. Derrida goes on to say of deconstruction that "wherever I have followed this investigative approach, it has been a question of showing that the system does not work, and that this dysfunction not only interrupts the system but itself accounts for the desire for system, which draws its *élan* from this very disadjoinment, or disjunction."

3. I point to blame not as a measure of all that happened on that platform—this is not an attempt to account for all of liberalism's ideas and ideals or a judgment of them—but as an effort to explain when, why, and how this new "-ism" came to be.

4. For the relationship of this doubling of culture to gender, see Siskin 1998a, 76–78. For its relationship to the viability of the enterprise of "Cultural Studies," see Siskin and Warner 2008, 102.

5. Keen 2004, 14. I cite De Quincey's essay from this anthology because Keen brilliantly juxtaposes it to a series of selections illustrating "The Nature of the Word, Literature."

6. I use the terms *Romantic* and *Romanticism* here to engage system's role in both the historical period (in Britain, the late eighteenth and early nineteenth centuries) and the tales literary study tells itself about the period. I have explored the relationship between these terms in Siskin 1988—which includes my account of Jerome McGann's effort to differentiate them (McGann 1985).

7. For my first effort to link system and Romanticism, see Siskin 2009b.

8. Hume's editor, L. A. Selby-Bigge, finds the addition of "Miracles" a mystery. Arguing that "they do not add anything to his general speculative position," he concludes that "their insertion in the Enquiry is due doubtless rather to other considerations than to a simple desire to illustrate or draw corollaries from the philosophical principles laid down in the original work." See Hume 1975, xix.

9. For the intentions of the *Review*'s founders, see Sullivan 1983, 139–142.

10. Byron, unpublished dedication to *Don Juan* 25–28 in McGann 1980–1993, V.4.

11. Review of *Don Juan* I and II in the *Edinburgh Monthly Review* (1819) cited in Haslett 1997, 119.

12. Jerome McGann, for example, reprised the Crabbe-as-exception argument in 1983.

13. For a very different take from mine on literature itself as a system—one that does not treat system as a genre with a history—see Guillen 1971.

14. For my first effort to link Wordsworth's authorial self and system, see Siskin 2006.

15. I use the term in a slightly different way than assimilating to a different culture, often the dominant one; at stake here is assimilation into culture itself as a new form of systematization.

16. For additional examples of that phenomenon, see Siskin 2005, 812–815.

17. For statistical evidence of relationship between "sublime" and "culture"—as well as a new history of the sublime based on those figures, see Algee-Hewitt 2008.

18. This title, with its sense of miscellany ("with a Few"), is clearly not a title for a complete system. Instead, it marks this initial attempt to put system into verse as a gesture *toward* system—an essayistic attempt at system as a goal still to be achieved. See the "Mixed Messages" section of chapter 3.

19. For an explanation of the importance of this strategy to the formation of modernity, including individual development, democracy, and professional power, see Siskin 1988, 63, 107–108.

20. *The Prelude* bears comparison, of course, to Rousseau's efforts as I have described them. But it is not confessions animated by journeys from system to system. It is an effort to shape a "life" into a developmental argument for the system project: the writing of a system of "men as they are men" that would yield neo-Newtonian laws of man.

21. Here is the Derrida quote in full: "If by 'system' is meant—and this is the minimal sense of the word—a sort of consequence, coherence and insistence—

a certain gathering together—there is an injunction to the system that I have never renounced, and never wished to. This can be seen in the recurrence of motifs and references from one text to another in my work, despite the differing occasions and pretexts—a recurrence that having reached a certain age, I find rather striking. What I have managed to write in the course of these past thirty years has been guided by a certain insistence that others may well find downright monotonous" (Derrida and Ferraris 2001, 3).

CHAPTER 7

1. Williams 1979, 154.

2. Blake 1970, 2.43.10.

3. Where Warner and I part ways with Williams is at his insistence that human "intention," arising out of "relations between ourselves," always precedes technological innovation." We argue for the need to rethink the binary of technodeterminism: "'mediation' embraces both the technological and the human—it does not discriminate, that is, against any particular form of agency" (Siskin and Warner 2010, 10). The point is not to counter Williams by privileging technology—and thus reinscribing the binary; rather, it is an exercise in transposing features of binaries in order to collapse them—and of valuing intention without sacralizing it as necessarily causal. As Kevin Kelly has recently argued, "Technology not only reveals our humanness, it is the way we are human." Technology is always already there when we, in fact, are.

A genre from the eighteenth century can help to make this point. In the Scottish conjectural histories—the narratives that underwrote the Enlightenment venture of the "human sciences"—man is a social being. The proof consisted basically of a single observation: no one is born alone. The child's mother is always there even if he or she is born in the middle of a desert. I would argue, however, that there is, in fact, something else there, something that our word for the experience of birth makes visible. We call it a "labor"—a form of work entailing the application of specific techniques and skills, in a more or less artful and methodical way, to be successful. From the regulation of breathing and contraction to the tools and procedures for cutting the umbilical cord, what is there with the mother and child is technology.

What we have at the moment of birth, then, is a human system, not the same one Douglas Engelbart had in mind, but it works in similar ways. The parts do talk to each other, and their conversation is of the genre of change I have been describing. It may seem strange to think of the baby's first cry as a formal feature of system, but that is its role in this conversation. Its function is to signal the success not only of this individual birth but of the cumulative, larger-scale technology at work: sexual reproduction as the mechanism for genetic change—the technology that, in a quite literal way, is the way we are, and will or will not be, human.

4. Raymond Williams's shorthand for the interrelated practices of writing, print, and silent reading (Williams 1983, 1–7).

5. This is the title of the main account, which is signed by "M. C." It was printed with prefatory letters under a title page that reads *An account of the Fair Intellectual-Club in Edinburgh: in a letter to a honorable member of an Athenian society there. By a young lady, the secretary of the club* (Edinburgh: J. McEuen and Company, 1720). Both the date—this may be the earliest record we have of this type of club for women in Britain—and the melodramatic frame of a secret exposed raise issues of authenticity. So far, however, my research in special collections in Scotland and consultations with other scholars have yielded no sign of a hoax. The strongest corroborative evidence, uncovered with the help of a typescript by Davis D. McElroy (1969), is the appearance in 1720 of the *Edinburgh Miscellany*, a collection of poems constituting the third book published by a group called the Athenian Society—the same group mentioned in the title of the *Account*. It includes poems by James Thomson and David Mallet; the latter's pastoral inscription of the volume, addressed to Joseph Mitchell, one of the Athenians, says of the female contributors, "Who the ladies are, scarce any one knows." Their group identity, however, was known. In his comments on contributors, the author of the Preface to the volume, a man identified as "W. C.," writes:

> As for the Ladies, who have generously contributed to make up this work, we are proud to declare, that, tho they have sent us few of their composures, they have sent nothing that is refuse. And therefore, while we publickly thank them for the Assistance already receiv'd, we beg they will continue to shine like the higher constellations among luminaries of a dimmer aspect. … we particularly thank the Fair Intellectual club for the poems they have been pleas'd to favor us with for publick use. And we presume the ingenious Readers of their performances will allow us to intreat them to send more to bespangle the second volume. (Anon. 1720, I.ii–iii)

See also McElroy's dissertation (1952) and the manuscript by Davis D. McElroy and L. A. McElroy, "Index to the Literary Clubs and Societies of Eighteenth Century Scotland" (1955).

I have identified two other specific, contemporary references to the club. Alexander Pennecuik's 1720 volume, *Streams from Helicon*, includes a poem entitled "To the Nine Muses, *Members of the fair Intellectual-Club*." The praise is extravagant:

> The Sister Arts to foreign Climates flown,
> Restor'd shall flowrish in our *Frigide Zone*:
> *Scotia's* cold Heaths become the Muses Seat,
> We'll once again be Wise, and once again be Great.

Awake *Apelles*, see the sacred *Nine*,
Have stoll'n thy Pencil, and thy Art divine;

. . . . .

Majestick *Nuns*, the World reform'd by you,
(Beauty perform'd, what Prophets could not do;)
With grateful *Anthems* at they Feet they'll bow.
When you fair Guides, let loose the Reins of Power,
You'll break the Cords of Sin, make the polluted pure,
By female virtue shall the Chains be broke,
And Beauties smiles, melt down the flinty Rock;
The glad'ning News shall spread to foreign lands,
That *Britain* rank in Sin, was sav'd by Virgins Hands.
(23–28, 59–67)

The other set of references, only a bit less enthusiastic, appear in the *Plain Dealer*, a London periodical, and are discussed in the main text.

6. I invoke Dawkins (1976) here for reasons of form rather than content; that is, my focus here is on the reversal of causal expectations rather than the specific consequences of that reversal. To continue the thought experiment, genres, of course, also replicate, adapting to new writing environments, as when Godwin, faced with jail, hid "things as they are" in a set of "adventures." Could Dawkins's eversion apply to the emergence of the "I" in Hays? Was that "I" an effect of generic adaptation? Of the genre of system surviving by embedding itself in other forms such as her *Appeal*?

7. The Ebola outbreak in West Africa in 2014–2015 illustrated what happens when this constraint is weakened by higher population density and increased mobility.

8. Just as the Web allows us to manipulate multiple identities today—stirring up anxieties about who someone "really" is—so writing, when it was the new technology, raised similar concerns. Who, for example, was the "M. C." who signed the *Account*? Those anxieties return in the present to haunt scholars like me, who must worry about the authenticity of these artifacts.

9. Far from having a special relationship to the real, argues Sadie Plant in her pioneering effort to construct a history of the relationship between women and technology, "face-to-face communication—the missionary position so beloved of Western man—is not at all the most direct of all possible ways to communicate" (Plant 1997, 143–144).

10. In Plant's terms, much of that scholarship might be said to be shot through with a nostalgia for the face-to-face that idealizes eighteenth-century social forms as models of democratic, civil society, forcing them to bear the suffocating weight of Habermasian desire.

11. For the specific role played by Scotland, Jacobitism, and the Act of Union in the formation of culture, see Siskin 1998a, 85–88.

12. In the seventeenth and eighteenth centuries, "Mrs." was prefixed to the name of an unmarried lady or girl in a manner equivalent to the modern use of "Miss." The use of the term here may also be connected to another use that died out during the eighteenth century: "Mrs." as an abbreviation of "mistress," a term for "a woman who rules, or has control." See *OED*.

13. I stumbled upon the *Account* by accident in the University of Glasgow Library in 1996, for it was largely ignored during the nineteenth and twentieth centuries. It has now resurfaced a bit in the twenty-first century. See Derya Gurses Tarbuck's description of the *Account* in *The Center and Clark Newsletter* (2008, 4–6). And, as noted in the main text, the Fair Intellectuals returned to public view in a play by Lucy Porter presented at the 2014 Edinburgh Festival.

14. Interview with John Eklund reported on Stanford University's "The Interface Project" website, which is no longer online.

15. Issues from 1724 were republished in two volumes in 1734. This quotation is from vol. 1, no. 46 (August 28, 1724).

16. Ironically, *Sync* magazine did not make a dime either. Launched by Ziff Davis in 2004, it was shut down in 2005.

17. The *Account* surfaced on a few occasions since then. In the bound volume 4 of *The Patrician*, the author of an article on "The Clubs of London" announces that "we can hardly bring ourselves to pass over" the opportunity to ridicule the club:

> It may be shrewdly expected that these *fair intellectuals*, if indeed they ever really existed, were fair after the inverted fashion of Macbeth's witches—"fair is foul, and foul is fair";—that they were silly pedants is beyond all question. (Burke 1847, 346–348)

More recently, Gale research reprinted the *Account* as an ECCO print edition in 2010. It has also been briefly noted in Clark 2000, 91, and Tarbuck 2008, 4–6.

CODA

1. Deutsch extends this argument about the alignment of knowledge and the world to the issue of control:

> People had dreamed for millennia of flying to the moon, but it was only with the advent of Newton's theories about the behaviour of invisible entities such as forces and momentum that they began to understand what was needed in order to go there. This increasingly intimate connection between explaining the world and controlling it is no accident, but is part of the deep structure of the world. (Deutsch 2011, 55)

There is a link back here to Bacon's sense of the relationship of knowledge to power as articulated by James C. Morrison:

> Bacon follows the tradition in defining philosophy as the search for the first causes and principles of all that is but at the same time radically transforms it by conceiving of the knowledge of causes as the knowledge of how to bring about or make something. That is, to know the cause of a thing is to know how to bring about that thing oneself: to know the cause of a given effect means to be able to produce that effect. (Morrison 1977, 591)

2. Whether one sees it as a consolidation and extension of earlier work on complexity, chaos theory, and autopoetic systems, or as a brilliant breakthrough in its own right, Wolfram's work in *A New Kind of Science* exemplifies one thread in the current reworking of the role of system.

3. See Wolfram's explanation of a cellular automaton and illustrations of the resulting patterns at http://mathworld.wolfram.com/CellularAutomaton.html, retrieved October 26, 2013.

4. That increase has been expressed most famously in Moore's Law, which claimed that the number of transistors on an affordable CPU would continue to double every two years. See Moore 1965.

5. See my tracking of this term from what I believe was its initial—and only other—use by John Rajchman: Siskin 1998a, 245, and Rajchman 1985, xiii.

6. For an example of "rooting" for a new deployment of system, see the work of Edgar Morin. In his work, we once again see system playing a key role in negotiating the relationship between the simple and the complex. *System*, he claims is "a root word for complexity," and "complexity is not a surface noise of the real, but is the very principle of the real." "The physical foundation of what we call reality," he claims, "is not simple, but is complex." For the "Paradigm of System" he advocates, the "whole-parts relations of system"—what I have been calling its scalability—consists of "*interactions*" best understood in terms of a dynamic notion of "organization" rather than the "static concept" of "structure." What's at stake for Morin here is that his "new notion of system breaks away from the classical ontology of object." What he wants from the "radicalness of this change" is "the innovation which it might bring about" through "psychophysical" "forms of action containing the awareness and control of the self. (Morin 1992, 137, 130, 122, 127, 126, 135, 138).

7. My point here is that Darwin's contribution has been less visible because it was made *in* writing. He reshaped knowledge by deploying the genre of system in its algorithmic mode.

8. By naming this type of delay, I am suggesting that sorting out the variations can be a useful tool in constructing a Baconian literary history. I thank William Warner for this particular name.

9. See Zenil's explanation of this phrase in a blogpost of June 10, 2011, http://www.mathrix.org/liquid/category/new-ideas.

10. Lloyd made this declaration in an interview with Kevin Kelly, whose definition of system as talking to itself I have frequently cited. See Kelly 2006, http://www.wired.com/2006/03/life-the-universe-and-everything.

REFERENCES

Abbott, Andrew Delano. 1988. *The System of Professions an Essay on the Division of Expert Labor*. Chicago: University of Chicago Press.

Abrams, M. H., and Stephen Greenblatt, eds. 2000. *The Norton Anthology of English Literature*. 7th ed. 2 vols. New York: Norton.

Algee-Hewitt, Mark. 2008. "The Afterlife of the Sublime: Toward a New History of Aesthetics in the Long Eighteenth Century." PhD diss., New York University.

Alsted, Johann Heinrich. 1989. *Encyclopaedia: Faksimile-Neudruck der Ausgabe Herborn 1630*. Stuttgart-Bad Cannstatt: Frommann-Holzboog.

Alstedius, Joannes Henricus. 1610a. *Artium liberalium, ac facultatum omnium systema mnemonicum*. Francof.

Alstedius, Joannes Henricus. 1610b. *Ioan: Henrici Alstedii Systema mnemonicum duplex*. Francof.

Annand, Alexander. 1832. *A Brief Outline of the Existing System for the Government of India: To Which is Annexed a Tabular Statement of Legislative Enactments from 1773 to 1826*. London: Saunders and Benning.

Arac, Jonathan. 2011. *Impure Worlds: The Institution of Literature in the Age of the Novel*. New York: Fordham University Press.

Arblaster, Anthony. 1984. *The Rise and Decline of Western Liberalism*. Oxford: Basil Blackwell.

Aristotle. 1930. *Physica*. Translated by Robert Purves Hardie and Russell Kerr Gaye. Oxford: Clarendon Press.

Armstrong, Nancy. 2005. *How Novels Think: The Limits of Individualism from 1719–1900*. New York: Columbia University Press.

Asiatic Society of Bengal. 1799. *Asiatick researches: or, Transactions of the Society instituted in Bengal: for inquiring into the history and antiquities, the arts, sciences, and literature, of Asia*. London: Printed for J. Sewell.

Austen, Jane. 1998. *Emma*. Edited by Terry Castle. Oxford: Oxford University Press.

Bacon, Francis. (1605) 1973. *The Advancement of Learning*. Edited by G.W. Kitchin. London: Dent.

Bacon, Francis. (1605/1627) 1974. *The Advancement of Learning and New Atlantis*. Edited by Arthur Johnston. Oxford: Clarendon Press.

Bacon, Francis. (1620) 1994. *Novum Organum with Other Parts of The Great Instauration*. Translated by P. Urbach and J. Gibson. Chicago: Open Court.

Bacon, Francis. (1620) 2000. "Franciscy de Verulamio, summi Angliæ cancellarij instauratio magna." In *Londini: Apud [B. Nortonium &] Ioannem Billium typographum Regium*. http://gateway.proquest.com/openurl?ctx_ver=Z39.88-2003&res_id=xri:eebo&rft_val_fmt=&rft_id=xri:eebo:image:21495.

Bacon, Francis. (1620) 2000. *The New Organon*. Edited by Lisa Jardine and Michael Silverthorne. Cambridge: Cambridge University Press.

Bacon, Francis. 1733. *The Philosophical Works of Francis Bacon*. 3 vols. Edited by Peter Shaw. London: Printed for J. J. and P. Knapton, D. Midwinter and A. Ward, A. Bettesworth and C. Hitch.

Bacon, Francis. 2008. *The Major Works: Including New Atlantis and the Essays*. Edited by Brian Vickers. Oxford: Oxford University Press.

Bankes, Thomas. 1788. *A New and Authentic System of Universal Geography, Antient and Modern*. London: Printed for C. Cooke.

Bannet, Eve Tavor. 2000. *The Domestic Revolution: Enlightenment Feminisms and the Novel*. Baltimore, MD: The Johns Hopkins University Press.

Barbauld, Anna Laetitia. 1810. "On the Origin and Progress of Novel-Writing." In *The British Novelists*, vol. 1, edited by Anna Laetitia Barbauld, 1–59. London: F. C. and J. Rivington.

Barney, Richard A. 1999. *Plots of Enlightenment: Education and the Novel in Eighteenth-Century England*. Stanford, CA: Stanford University Press.

Barry, Andrew, and Georgina Born, eds. 2013. *Interdisciplinarity: Reconfigurations of the Social and Natural Sciences*. New York: Routledge.

Barry, Peter. 1997. "Romanticism, Regendering, and the Dearth of Poetry." *Cambridge Quarterly* 26 (3): 205–218.

Baumgarten, Alexander Gottlieb. 1750–1758. *Aesthetica*. Frankfurt an der Oder: I. C. Kleyb.

Bayle, Pierre. 1991. *Historical and Critical Dictionary: Selections*. Translated by Richard H. Popkin. Indianapolis, IN: Hackett.

Bazerman, Charles. 1988. *Shaping Written Knowledge: The Genre and Activity of the Experimental Article in Science*. Madison: University of Wisconsin Press.

Beer, John. 1995. "Fragmentations and Ironies." In *Questioning Romanticism*, edited by John Beer, 234–264. Baltimore, MD: The Johns Hopkins University Press.

Behn, Aphra. 1992–1995. *The Works of Aphra Behn*. 7 vols. Edited by Janet Todd. London: Pickering.

Belcher, William. 1798. *Intellectual electricity, novum organum of vision, and grand mystic secret ...: being an experimental and practical system of the passions, metaphysics and religion, really genuine: accompanied with appropriate extracts from Sir Isaac Newton, Dr. Hartley, Beddoes, and others: with medical observations rising out of the subject/by a rational mystic*. London: Lee & Hurst.

Bender, John. 1992. "A New History of the Enlightenment?" In *The Profession of Eighteenth-Century Literature: Reflections on an Institution*, edited by Leo Damrosch, 62–83. Madison: University of Wisconsin Press.

Bender, John, and Michael Marrinan. 2010. *The Culture of Diagram*. Stanford, CA: Stanford University Press.

Berlin, Isaiah. 1999. *The Roots of Romanticism*. Edited by Henry Hardy. London: Chatto and Windus.

Blackmore, Richard. 1712. *Creation. A Philosophical Poem. Demonstrating the Existence and Providence of a God. In Seven Books*. 3rd ed. London: Printed for S. Buckley & J. Tonson.

Blackwell, Richard J. 1991. *Galileo, Bellarmine, and the Bible: Including a Translation of Foscarini's Letter on the Motion of the Earth*. Notre Dame, IN: University of Notre Dame Press.

Blake, William. 1966. *The Complete Writings of William Blake*. Edited by Geoffrey Keynes. Oxford: Oxford University Press.

Blake, William. 1970. *The Poetry and Prose of William Blake*. Edited by David V. Erdman. New York: Doubleday.

Bolter, Jay David, and Richard Grusin. 1999. *Remediation: Understanding New Media*. Cambridge, MA: MIT Press.

Booth, Benjamin. 1789. *A complete system of book-keeping, by an improved mode of double-entry: comprising a regular series of transactions, as they have actually occurred in business ... To which are added, a new method of stating factorage accounts ... a concise ... view of the exchanges between all the principal trading cities of Europe*. London: Couchman & Fry.

Born, Daniel. 1995. *The Birth of Liberal Guilt in the English Novel: Charles Dickens to H. G. Wells*. Chapel Hill: University of North Carolina Press.

Boswell, James. 1984. *A Journey to the Western Islands Scotland, and The Journal of a Tour to the Hebrides*. Harmondsworth, England: Penguin.

Boyle, Robert. 1660. *New experiments physico-mechanicall, touching the spring of the air, and its effects: (made, for the most part, in a new pneumatical engine) written by way of letter to the Right Honorable Charles Lord Vicount of Dungarvan, eldest son to the Earl of Corke*. Oxford: Printed by Henry Hall for Thomas Robinson.

Boyle, Robert. 1661. *Certain physiological essays and other tracts written at distant times, and on several occasions by the honourable Robert Boyle; wherein some of the tracts are enlarged by experiments and the work is increased by the addition of a discourse about the absolute rest in bodies*. London: Printed for Henry Herringman.

Bredvold, Louis. 1962. *The Natural History of Sensibility*. Detroit, MI: Wayne State University Press.

Brigham, Linda C. 1996. "Observing the Observers of Systems and Environments." http://www.electronicbookreview.com/thread/criticalecologies/internal, accessed April 12, 2012.

Brown, Gerard Baldwin, A. Blyth Webster, and Eric Hyde Lord Sexton. 1903. *The Arts in Early England*. 6 vols. London: J. Murray.

Brown, Homer Obed. 1997. *Institutions of the English Novel from Defoe to Scott*. Philadelphia: University of Pennsylvania Press.

Brown, Marshall. 1991. *Preromanticism*. Stanford, CA: Stanford University Press.

Brown, Marshall. 1997. "Romanticism and Enlightenment." In *The Cambridge Companion to British Romanticism*, edited by Stuart Curran, 195–219. Cambridge: Cambridge University Press.

Buchan, James. 2003. *Crowded with Genius: The Scottish Enlightenment: Edinburgh's Moment of the Mind*. New York: HarperCollins.

Bullock, Alan, and Maurice Shock, eds. 1956. *The Liberal Tradition from Fox to Keynes*. Oxford: Clarendon Press.

Burgess, Miranda. 2000. *British Fiction and the Production of Social Order, 1740–1830*. Cambridge: Cambridge University Press.

Burke, Bernard, and John Burke, eds. 1847. *The Patrician*. London: E. Churton.

Burke, Edmund. (1757/1759) 1958. *A Philosophical Enquiry into the Origin of Our Ideas of the Sublime and Beautiful*. Edited by James T. Boulton. Notre Dame: University of Notre Dame Press.

Burke, Peter. 2000. *A Social History of Knowledge: From Gutenberg to Diderot*. Malden, MA: Polity Press.

Burke, Peter. 2012. *A Social History of Knowledge II: From the Encyclopaedia to Wikipedia*. Malden, MA: Polity Press.

Burrow, J. W. 1988. *Whigs and Liberals: Continuity and Change in English Political Thought: The Carlyle Lectures 1985*. Oxford: Clarendon Press.

Burton, John, and George Stubbs. 1751. *An essay towards a complete new system of midwifry, theoretical and practical … Interspersed with several new improvements; … illustrated with … eighteen copper-plates. In four parts*. London: Printed for James Hodges.

Butler, Marilyn. 1975. *Jane Austen and the War of Ideas*. Oxford: Clarendon Press.

Butt, John. 1979. *The Mid-Eighteenth Century*. Edited by John Buxton and Norman Davis. Oxford: Clarendon Press.

C., M. 1720. *An account of the Fair Intellectual-Club in Edinburgh: In a letter to a honourable member of an Athenian society there. By a young lady, the secretary of the club.* Edinburgh: J. McEuen and Company.

Campbell, George. 1762. *A dissertation on miracles: containing an examination of the principles advanced by David Hume, Esq; in an Essay on miracles*. Edinburgh: A. Kincaid & J. Bell.

Canning, George. 1817. *Substance of the speech of … George Canning … on sir M. W. Ridley's motion for reducing the number of the lords of the Admiralty*. London: Printed for John Murray.

Carruthers, Mary J. 1990. *The Book of Memory: A Study of Memory in Medieval Culture*. Cambridge: Cambridge University Press.

Carter, John, and Ronald McRae. 1997. *The Routledge History of Literature in English: Britain and Ireland*. London: Routledge.

Casaubon, Isaac, and Jacques Davy Du Perron. 1612. *The answere of master Isaac Casaubon to the epistle of … cardinall Peron*. London: F. Kyngston for W. Aspley.

Cassirer, Ernst. 1950. *The Problem of Knowledge; Philosophy, Science, and History since Hegel*. Translated by William H. Woglam, and Charles W. Hendel. New Haven, CT: Yale University Press.

Cassirer, Ernst. 1955. *The Philosophy of Enlightenment*. Translated by F. C. A. Koelin and J. P. Pettegrove. Boston: Beacon Press.

Chambers, Ephraim. 1728. *Cyclopædia: or, An universal dictionary of arts and sciences: containing the definitions of the terms, and accounts of the things signify'd thereby, in the several arts, both liberal and mechanical, and the several sciences, human and divine: the figures, kinds, properties, productions, preparations, and uses, of things natural and artificial: the rise, progress, and state of things ecclesiastical, civil, military, and commercial: with the several systems, sects, opinions, &c. among philosophers, divines, mathematicians, physicians, antiquaries, criticks, &c: the whole intended as a course of antient and modern learning*. 2 vols. London: Printed for James and John Knapton.

Chambers, Ephraim. 1738. *Cyclopaedia, or, An universal dictionary of arts and sciences … The whole intended as a course of antient and modern learning, extracted from the best authors, dictionaries, journals, memoirs, transactions, ephemerides, &c. in several languages*, 2nd ed. 2 vols. London: D. Midwinter.

Clark, David L., and Donald C. Goellnicht, eds. 1994. *New Romanticisms: Theory and Critical Practice.* Toronto: University of Toronto Press.

Clark, Malcolm. 1971. *Logic and System: A Study of the Transition from "Vorstellung" to Thought in the Philosophy of Hegel.* The Hague: Nijhoff.

Clark, Peter. 2000. *British Clubs and Societies, 1580–1800: The Origins of an Associational World.* Oxford: Clarendon Press.

Clark, William. 2006. *Academic Charisma and the Origins of the Research University.* Chicago: University of Chicago Press.

Cohen, Adam. 1997. "Battle of the Binge." *Time*, September 8.

Cohen, I. Bernard. 1980. *The Newtonian Revolution: With Illustrations of the Transformation of Scientific Ideas.* Cambridge: Cambridge University Press.

Cohen, Ralph. 1982. "Review of *The Creative Imagination: Enlightenment to Romanticism,* by James Engell." *Criticism* 24 (2): 174–180.

Cohen, Ralph. 1986. "History and Genre." *New Literary History* 17: 203–218.

Cohen, Ralph. 1993. "Generating Literary Histories." In *New Historical Literary Study: Essays on Reproducing Texts, Representing History*, edited by Jeffrey N. Cox and Larry J. Reynolds, 39–53. Princeton, NJ: Princeton University Press.

Cohen, Ralph. 2008a. "Farewell." *New Literary History* 39 (3): i.

Cohen, Ralph. 2008b. "Introduction." *New Literary History* 39 (3): vii–xx.

Cohen, Ralph, and Michael S. Roth, eds. 1995. *History and … :Histories within the Human Sciences.* Charlottesville: University of Virginia Press.

Coleridge, Samuel Taylor. 1971–2001. *The Collected Works of Samuel Taylor Coleridge.* Edited by Kathleen Coburn. Princeton, NJ: Princeton University Press.

Coleridge, Samuel Taylor. 1995. *Shorter Works and Fragments.* 2 vols. Edited by H. J. Jackson, and J. R. de J. Jackson. Princeton, NJ: Princeton University Press.

Coleridge, Samuel Taylor, Henry Nelson Coleridge, John Taylor Coleridge, and Carl Woodring. 1990. *Table Talk.* 2 vols. Princeton, NJ: Princeton University Press.

Collins, William. 1979. *The Works of William Collins.* Edited by Richard Wendorf and Charles Ryskamp. Oxford: Clarendon Press.

Condillac, Abbé de. (1749) 1982. "A Treatise on Systems." In *Philosophical Writings of Etienne Bonnot, Abbe de Condillac.* Hillsdale, NJ: Erlbaum.

Condillac, Abbé de. 1947–1951. "Dictionnaire des synonymes." In *Oeuvres philosophiques de Condillac.* Edited by Georges LeRoy. Paris: Presses universitaires de France.

Conger, Syndy M. 1994. *Mary Wollstonecraft and the Language of Sensibility.* Rutherford, NJ: Fairleigh Dickinson University Press.

Court, Franklin E. 1992. *Institutionalizing English Literature: The Culture and Politics of Literary Study, 1750–1900*. Stanford, CA: Stanford University Press.

Courthope, William John. 1885. *The Liberal Movement in English Literature*. London: John Murray.

Cowan, Brian William. 2005. *The Social Life of Coffee: The Emergence of the British Coffeehouse*. New Haven, CT: Yale University Press.

Crane, Susan A. 2000. *Museums and Memory*. Stanford, CA: Stanford University Press.

Crawford, Robert. 1992. *Devolving English Literature*. Oxford: Clarendon Press.

Cronin, Richard. 2003. "Review of *The Historical Austen*, by William H. Galperin." *Wordsworth Circle* 34 (4): 192–194.

Cruickshanks, Eveline. 1988. *The Jacobite Challenge*. Edited by Jeremy Black. Edinburgh: John Donald.

Curran, Stuart, ed. 2010. *The Cambridge Companion to British Romanticism*, 2nd ed. Cambridge: Cambridge University Press.

Dards, Mrs. 1800. *A catalogue of shell-work, &c. … Consisting of a great variety of beautiful objects, equal to nature, minutely described; comprehending a new System, which will be highly gratifying to every lover of natural history.* London: H. Fry.

Darwin, Charles. 1909. *The Foundations of the Origin of Species: Two Essays Written in 1842 and 1844*. Edited by Francis Darwin. Cambridge: Cambridge University Press.

Davis, Leith, Ian Duncan, and Janet Sorensen. 2004. *Scotland and the Borders of Romanticism*. Cambridge: Cambridge University Press.

Dawkins, Richard. 1976. *The Selfish Gene*. Oxford: Oxford University Press.

Day, Aidan. 1996. *Romanticism*. New York: Routledge.

de Bolla, Peter. 2013. *The Architecture of Concepts: The Historical Formation of Human Rights*. New York: Fordham University Press.

Delany, Samuel R. 1980. "Generic Protocols: Science Fiction and Mundane." In *The Technological Imagination: Theories and Fictions*, edited by Teresa De Lauretis, Andreas Huyssen, and Kathleen Woodward, 175–193. Madison, WI: Coda Press.

DeMaria, Robert Jr., ed. 1996. *British Literature 1640–1789: An Anthology*. Oxford: Blackwell.

Dennett, Daniel C. 1995. *Darwin's Dangerous Idea: Evolution and the Meanings of Life*. New York: Touchstone.

Derrida, Jacques, and Maurizio Ferraris. 2001. *A Taste for the Secret*. Edited by Giacomo Donis and David Webb. Translated by Giacomo Donis. Malden, MA: Polity.

Descartes, René. 1994. *A Discourse on Method; Meditations on the first Philosophy; Principles of Philosophy*. Edited by Tom Sorell. Translated by John Veitch. London: Everyman.

Deutsch, David. 1997. *The Fabric of Reality*. London: Penguin.

Deutsch, David. 2011. *The Beginning of Infinity: Explanations That Transform the World*. London: Penguin.

Deutsch, David. 2012a. "Constructor Theory." *arXiv*: eprint arXiv:1210.7439.

Deutsch, David. 2012b. "Creative Blocks: The Very Laws of Physics Imply That Artificial Intelligence Must Be Possible. What's Holding Us Up?" *Aeon*, October 3. https://aeon.co/essays/how-close-are-we-to-creating-artificial-intelligence.

Dick, Philip K. 1968. *Do Androids Dream of Electric Sheep?* New York: Del Rey.

Dickinson, H.T. 1977. *Liberty and Property: Political Ideology in Eighteenth-Century Britain*. New York: Holmes & Meier.

Donoghue, William. 2002. *Enlightenment Fiction in England, France, and America*. Gainesville: University Press of Florida.

Doody, Margaret Anne. 1996. *The True Story of the Novel*. Rutgers: Rutgers University Press.

Dowling, William C. 1992. "Ideology and the Flight from History in Eighteenth-Century Poetry." In *The Profession of Eighteenth-Century Literature: Reflections on an Institution*, edited by Leo Damrosch, 135–153. Madison: University of Wisconsin Press.

Duff, David. 1998. "From Revolution to Romanticism: The Historical Context to 1800." In *A Companion to Romanticism*, edited by Duncan Wu, 23–34. Malden, MA: Blackwell.

Duncan, James. 1816. *A new introduction to stenography, or short-hand writing; being an attempt to facilitate the acquisition of Dr. Mavor's standard system, by a more simple elucidation of its elements*. 2nd ed. 8 vols. Glasgow: Printed for J. Duncan.

Dwyer, John. 1987. *Virtuous Discourse: Sensibility and Community in Late Eighteenth-Century Scotland*. Edinburgh: J. Donald.

Dylan, Bob. 1973. *Writings and Drawings*. New York: Knopf.

Dyson, Freeman. 2006. "Make Me a Hipporoo." *New Scientist* 189 (2538): 36–39.

Eccleshall, Robert. 1986. *British Liberalism: Liberal Thought from the 1640s to 1980s*. Edited by Bernard Crick. London: Longman.

The Edinburgh Miscellany: Consisting of Original Poem, Translations, Etc., by Various Hands. Vol. 1. 1720. Edinburgh: J. McEuen and Company.

Edwards, Paul N. 2010. *A Vast Machine: Computer Models, Climate Data, and the Politics of Global Warming*. Cambridge, MA: MIT Press.

Edwards, Paul N., Steven J. Jackson, Geoffrey C. Bowker, and Cory P. Knobel. 2007. *Understanding Infrastructure: Dynamics, Tensions, and Design: Report of a Workshop on "History & Theory of Infrastructure: Lessons for New Scientific Cyberinfrastructures."* January. NSF Grant 0630263.

Einstein, Albert. 1954. *Ideas and Opinions. Based on Mein Weltbild.* New York: Crown.

Emanuel, Rahm. 1997. "Convocation Address." School of Speech, Northwestern University, June 21.

Encyclopaedia Britannica. 1777–1784. Edited by James Tytler. 2nd ed. 10 vols. Edinburgh.

Encyclopaedia Britannica, or a dictionary of arts, sciences and miscellaneous literature. … 1797. 3rd ed. 18 vols. Edinburgh: Printed for A. Bell, and C. MacFarquhar.

Encyclopaedia Britannica: or, a dictionary of arts and sciences compiled upon a new plan in which the different sciences and arts are digested into distinct treatises or systems and the various technical terms, &c. are explained as they occur in the order of the alphabet / by a society of gentlemen in Scotland. 1771. 3 vols. Edinburgh: Printed for A. Bell and C. Macfarquhar.

"English Short Title Catalogue." Bibliographic database. Eureka.

Engelbart, D. C. 1962. *Augmenting Human Intellect: A Conceptual Framework.* Washington, DC: Air Force Office of Scientific Research.

An Entire New System of Mercantile Calculation, By the Use of Universal Arbiter Numbers. Introduced by an Elementary Description of, and Commercial and Political Reflections on Universal Trade. Illustrated and Exemplified by the Elements of the Chain Rule of Three, the Nature of the Exchanges, and of all Charges and Contingencies on Goods; Which are also Reduced to a Plain and Concise System, Intirely New and Universal. 1795. London: Printed for Leigh and Sotheby.

Erickson, Lee. 1996. *The Economy of Literary Form: English Literature and the Industrialization of Publishing 1800–1850.* Baltimore, MD: The Johns Hopkins University Press.

Erickson, Lee. 2002. "'Unboastful Bard': Originally Anonymous English Romantic Poetry Book Publication, 1770–1835." *New Literary History* 33 (2): 247–278.

The Eton System of Education Vindicated: And its Capabilities of Improvement Considered, in Reply to Some Recent Publications. 1834. London: Printed for J. G. & F. Rivington.

Ewing, Thomas. 1816. *A system of geography, for the use of schools and private students, on a new and easy plan: in which the European boundaries are stated as settled by the Treaty of Paris and Congress of Vienna: with an account of the solar system, and a variety of problems to be solved by the terrestrial and celestial globes.* Edinburgh: Oliver and Boyd.

Fairer, David. 1997. "Organizing Verse: Burke's Reflections and Eighteenth-Century Poetry." *Romanticism* 3 (1): 1–19.

Favret, Mary. 2003. "Review of *The Historical Austen*, by William H. Galperin." *Romantic Circles Reviews* 7 (1): 10, http://www.rc.umd.edu/reviews-blog/william-h-galperin-historical-austen.

Feingold, Mordechai. 2004. *The Newtonian Moment: Isaac Newton and the Making of Modern Culture*. New York: Oxford University Press.

Ferguson, James. 1835. *A complete system of mental arithmetic, embracing every variety of commercial calculation necessary for the counting-house and shop*. Alnwick: Printed by M. Smith.

Fielding, Henry. (1741–1742) 1961. *Joseph Andrews and Shamela*. Edited by Martin Battestin. Boston: Houghton Mifflin.

Fielding, Henry. 1742. *The History of the Adventures of Joseph Andrews, and of His Friend Mr. Abraham Adams. … In two volumes*. London: Printed for A. Millar.

Fielding, Henry. (1742) 1999. *Joseph Andrews and Shamela*. Edited by Judith Hawley. London: Penguin.

Fielding, Henry. (1749) 1996. *Tom Jones*. Edited by John Bender and Simon Stern. Oxford: Oxford University Press.

Fielding, Sarah. 1749. *The Governess: or, Little Female Academy. Being the History of Mrs. Teachum, and Her Nine Girls: With Their Nine Days Amusement: Calculated for the Entertainment and Instruction of Young Ladies in Their Education*. London: A. Millar.

Findlen, Paula. 1994. *Possessing Nature: Museums, Collecting, and Scientific Culture in Early Modern Italy*. Berkeley: University of California Press.

Firestone, David. 2000. "Ex-Radical Cites 'System' in His Defense." *New York Times*, March 23.

Fischer, David Hackett. 1996. *The Great Wave: Price Revolutions and the Rhythm of History*. New York: Oxford University Press.

Fitzgibbons, Athol. 1995. *Adam Smith's System of Liberty, Wealth, and Virtue: The Moral and Political Foundations of The Wealth of Nations*. Oxford: Clarendon Press.

Fletcher, Andrew. 1704. *An account of a conversation concerning a right regulation of governments, for the common good of mankind; in a letter to the Marquess of Montrose, the Earls of Rothes, Roxburgh, and Haddington*. London: A. Baldwin.

Fletcher, Winston. 2000. "'When Three Hit a Sticky Wicket': Review of *The Tipping Point*, by Malcolm Gladwell." *Times Higher Education Supplement*, September 29.

Folger Collective on Early Women Critics, eds. 1995. *Women Critics 1660–1820: An Anthology*. Bloomington: Indiana University Press.

Fontenelle, Bernard Le Bovier de. 1801. *A week's conversation on the plurality of worlds. Translated by Mrs. A. Behn … [and others]. To which is added, Mr. Addison's Defence on the Newtonian philosophy*. 7th ed. with considerable improvements. London.

Fontenelle, Bernard Le Bovier de. 1990. *Entretiens sur la pluralité des mondes: Conversations on the Plurality of Worlds*. Translated by H. A. Hargreaves. Berkeley: University of California Press.

Forman, Simon. 1591. *The groundes of the longitude*. London: T. Dawson.

Foster, John. 1974. *Class Struggle and the Industrial Revolution: Early Industrial Capitalism in Three English Towns*. New York: St. Martin's Press.

Foucault, Michel. 1984. "What Is Enlightenment?" In *The Foucault Reader*, edited by Paul Rabinow, 32–50. New York: Pantheon Books.

Foucault, Michel. 1988. "The Functions of Literature." In *Politics, Philosophy, Culture: Interviews and Other Writings, 1977–1984*, edited by Lawrence D. Kritzman, 307–313. New York: Routledge.

Foucault, Michel. 1998. *Aesthetics, Method, and Epistemology: Essential Works of Foucault, 1954–1984*. Edited by James D. Faubion. New York: New Press.

Foucault, Michel. 2007. *The Politics of Truth*. Edited by Sylvère Lotringer. Translated by Lysa Hochroth and Catherine Porter. New York: Semiotext(e).

Freedman, Joseph S. 1997. "The Career and Writings of Bartholomew Keckermann (d. 1609)." *Proceedings of the American Philosophical Society* 141 (3): 305–364.

Freeholder. 1784. "The source of the evil or, the system displayed. Addressed to the gentry, yoemanry [*sic*], freeholders, and electors of England. By a freeholder. Letter I." London: Printed and sold by all the booksellers in town and country. Eighteenth Century Collections Online (CW107959238).

Frick, Arne, Andreas Ludwig, and Heiko Mehldau. 1994. "A Fast Adaptive Layout Algorithm for Undirected Graphs (Extended Abstract and System Demonstration)." In *Lecture Notes in Computer Science* 894 (Proceedings of Graph Drawing '94), edited by Roberto Tamassia, and Ioannis G. Tollis, 388–403. New York: Springer.

Fruchterman, Thomas M. J., and Edward M. Reingold. 1991. "Graph Drawing by Force-Directed Placement." *Software: Practice and Experience* 21 (11): 1129–1164.

Frye, Northrop. 1956. "Towards Defining an Age of Sensibility." *English Literary History* 23 (2): 144–152.

Frye, Northrop. 1990–1991. "Varieties of Eighteenth Century Sensibility." *Eighteenth-Century Studies* 24 (2): 157–172.

Furetière, Antoine, Delaune, Ganeau, and Gandouin. 1704. *Dictionnaire universel francois et latin: contenant la signification et la definition tant des mots de l'une & de l'autre Langue, avec leurs différens usages, que des termes propres de chaque Etat & de chaque Profession. La Description de toutes les chose naturelles & artificielles; leurs figures, leurs espèces, leurs propriétés. L'Explication de tout ce que renferment les Sciences & les Arts, foit Libéraux, foit Méchaniques*. Paris: Chez la Veuve Delaune, La Veuve Ganeau, rue S. Jacques. Gandouin, Quai dea Agustins.

Galilei, Galileo (1610) 1989. *Sidereus nuncius, or, The Sidereal Messenger*. Edited and translated by Albert Van Helden. Chicago: University of Chicago Press.

Galilei, Galileo. (1632) 1953. *Dialogue Concerning the Two Chief World Systems—Ptolemaic & Copernican*. Translated by Stillman Drake. Berkeley: University of California Press.

Galilei, Galileo. 2008. *The Essential Galileo*. Edited and translated by Maurice A. Finocchiaro. Indianapolis, IN: Hackett.

Gansner, Emden, et al. 2009. "Putting Recommendations on the Map—Visualizing Clusters and Relations." In *Proceedings of the Third ACM Conference on Recommender Systems*, 345–348. New York: Association for Computing Machinery.

Garside, Peter, James Raven, and Rainer Schöwerling. 2000. *The English Novel, 1770–1829: A Bibliographical Survey of Prose Fiction Published in the British Isles*, 2 vols. Oxford: Oxford University Press.

George-Parkin, Hilary. 2013. "Meet the New Fashion Mag That's Already Rubbing Condé Nast the Wrong Way." *Styleite*.

Gerrard, Christine, and David Fairer, eds. 1999. *Eighteenth-Century Poetry: An Annotated Anthology*. Malden, MA: Blackwell.

Gibbon, Edward. 1776–1778. *The history of the decline and fall of the Roman Empire*. 6 vols. London: Printed for A. Strahan, and T. Cadell.

Gilmartin, Kevin. 1996. *Print Politics: The Press and Radical Opposition in Early Nineteenth-Century England*. Cambridge: Cambridge University Press.

Gilmartin, Kevin. 1997. "William Cobbett and the Politics of System." *Romantic Circles Praxis: Romanticism and Conspiracy*, http://www.rc.umd.edu/praxis/conspiracy/gilmartin/kg2.html.

Gladwell, Malcolm. 2002. *The Tipping Point: How Little Things Can Make a Big Difference*. London: Abacus.

Gladwin, Francis. 1790. *A compendious system of Bengal revenue accounts*, 2nd ed. Calcutta: Printed by Manuel Cantopher.

Goclenius, Rudolph, Bartholomäus Keckermann, and Johann Stöckle. n.d. *Resolutio systematis logici majoris in tabellas pleniores, quam quæ antehac fuerunt*. Francofurti: Apud Ioannem Stöckle.

Godwin, William. 1793. *An Enquiry Concerning Political Justice, and Its Influence on General Virtue and Happiness*. 2 vols. London: Printed for G. G. J. and J. Robinson.

Godwin, William. (1793) 1993. "An Enquiry Concerning Political Justice." In *Political and Philosophical Writings of William Godwin*, vol. 3, edited by Mark Philp. London: William Pickering.

Godwin, William. (1794) 1988. *Things as They Are or The Adventures of Caleb Williams*. Edited by Maurice Hindle. London: Penguin.

Godwin, William. 1797. *The Enquirer: Reflections on Education, Manners, and Literature. In a Series of Essays*. London: Printed for G. G. and J. Robinson, Paternoster-Row.

Godwin, William. 1820. *Of Population: An Enquiry Concerning the Power of Increase in the Numbers of Mankind, Being an Answer to Mr. Malthus's Essay on that Subject*. London: Printed for Longman, Hurst, Rees, Orme, and Brown.

Godwin, William. (1820) 1976. "Of Population: An Enquiry concerning the Power of Increase in the Numbers of Mankind." In *An Essay on the Principle of Population*, edited by Philip Appleman. New York: Norton.

Godwin, William. 1832. *Fleetwood: or, The New Man of Feeling*. London: R. Bentley.

Gordon, Daniel, ed. 2001. *Postmodernism and the Enlightenment: New Perspectives in Eighteenth-Century French Intellectual History*. London: Routledge.

Gossman, Lionel. 1990. *Between History and Literature*. Cambridge, MA: Harvard University Press.

Goyder, D. G. 1825. *A manual of the system of instruction pursued at the infant school, Meadow Street, Bristol: illustrated by appropriate engravings*, 4th ed. London: Longman, Hurst, Rees, Orme, Brown and Green.

Grady, Thomas, and Bernard Dornin. 1799. *An impartial view, of the causes leading this country to the necessity of an union; in which the two leading characters of the state are contrasted; and in which is contained, a history of the rise and progress of Orange Men; a reply to Cease your funning, and Mr. Jebb*, 3rd ed. Dublin: Printed for B. Dornin.

Granger, James. 1769. *A biographical history of England: from Egbert the Great to the Revolution: consisting of characters disposed in different classes, and adapted to a methodical catalogue of engraved British heads. Intended as an essay towards reducing our biography to system, and a help to the knowledge of portraits … . With a preface, shewing the utility of a collection of engraved portraits to supply the defect, and answer the various purposes, of medals*. Vol. 1. London: Printed for T. Davies.

Granger, James. 1779. *A Biographical History of England, from Egbert the Great to the Revolution: consisting of* CHARACTERS *disposed in different* CLASSES, *and adapted to a* METHODICAL CATALOGUE *of Engraved* BRITISH HEADS. INTENDED *as An Essay towards reducing our* BIOGRAPHY *to* SYSTEM, *and a* HELP *to the Knowledge of* PORTRAITS. INTERSPERSED WITH *variety of* ANECDOTES, *and* MEMOIRS *of a great number of* PERSONS, *not to be found in any other Biographical Work. With a* PREFACE, *shewing the Utility of a Collection of* ENGRAVED PORTRAITS *to supply the Defect, and answer the various Purposes of* MEDALS, 3rd ed. London: J. Rivington and Sons, B. Law, J. Robson, G. Robinson, T. Cadell, T. Evans, R. Baldwin, J. Nicholl, W. Oteridge, and Fielding and Walker.

Grant, Edward. 2007. *A History of Natural Philosophy: From the Ancient World to the Nineteenth Century*. Cambridge: Cambridge University Press.

Grant, Ruth W. 1987. *John Locke's Liberalism*. Chicago: University of Chicago Press.

Gray, John. 1995. *Liberalism*. 2nd ed. Edited by Frank Parkin. Minneapolis: University of Minnesota Press.

Gray, Thomas. 1935. *The Correspondence of Thomas Gray*. 3 vols. Edited by Paget Toynbee, and Leonard Whibley. Oxford: Clarendon Press.

Gray, Thomas, William Collins, and Oliver Goldsmith. 1969. *The Poems of Thomas Gray, William Collins, Oliver Goldsmith*. Edited by Roger Lonsdale. London: Longmans, Green & Co.

Greenwald, Ted. 2007. "The Dark World of Ridley Scott." *Wired*, October, 178–185.

Guillen, Claudio. 1971. *Literature as System: Essays toward the Theory of Literary History*. Princeton, NJ: Princeton University Press.

Hamilton, Sir William. 1859–1860. *Lectures on Metaphysics and Logic*. 4 vols. Edited by H. L. Mansel and J. Veitch. Edinburgh: William Blackwood & Sons.

Haney, John Louis. (1904) 1970. *Early Reviews of English Poets*. New York: Burt Franklin.

Harding, Sandra G. 1991. *Whose Science? Whose Knowledge? Thinking from Women's Lives*. Ithaca, NY: Cornell University Press.

Harkness, Deborah E. 2007. *The Jewel House: Elizabethan London and the Scientific Revolution*. New Haven, CT: Yale University Press.

Haslett, Moyra. 1997. *Byron's Don Juan and the Don Juan Legend*. Oxford: Clarendon Press.

Hayes, Julie C. 1998. "Fictions of Enlightenment: Sontag, Süskind, Norfolk, Kurzweil." In *Questioning History: The Postmodern Turn to the Eighteenth Century*, edited by Greg Clingham, 21–36. Lewisburg, PA: Bucknell University Press.

Hays, Mary. (1798) 1974. *Appeal to the Men of Great Britain in Behalf of Women*. Edited by Gina Luria. New York: Garland.

Hays, Mary. 2009. *Memoirs of Emma Courtney*. Edited by Eleanor Rose Ty. New York: Oxford University Press.

Hays, Mary, J. Johnson, and John Bell. 1798. *Appeal to the Men of Great Britain in behalf of Women*. London: Printed for J. Johnson, St. Paul's Church-Yard; and J. Bell, Oxford-Street.

Hazlitt, William. 1825. *The Spirit of the Age, or, Contemporary Portraits*, 2nd ed. London: Printed for Henry Colburn.

Hilles, Frederick W. 1965. *From Sensibility to Romanticism: Essays Presented to Frederick A. Pottle*. Edited by Harold Bloom. New York: Oxford University Press.

Holland, John H. 2000. *Emergence: From Chaos to Order*. Oxford: Oxford University Press.

Holt, Francis Ludlow. 1820. *A system of the shipping and navigation laws of Great Britain: and of the laws relative to merchant ships and seamen, and maritime contracts ... to which is added, an appendix of acts of Parliament, forms, &c.* 2 vols. London: Printed for J. Butterworth.

Horkheimer, Max, and Theodor W. Adorno. 1996. *Dialectic of Enlightenment*. Translated by John Cumming. New York: Continuum.

Hoyle, Ben. 2009. "It's 1759 and All That ... or the History You Never Learnt at School." *Times*, January 15.

Hume, David. (1758/1777) 1902/1975. *Essays and Treatises on Several Subjects*. London: Printed for T. Cadell.

Hume, David. 1975. *Enquiries Concerning the Human Understanding and Concerning the Principles of Morals*, 3rd ed. Edited by L. A. Selby-Bigge, and P. H. Nidditch. Oxford: Clarendon Press.

Hunt, Leigh. (1828) 1976. *Lord Byron and Some of His Contemporaries*. New York: Georg Olms.

Hutchings, William, and W. B. Ruddick, eds. 1993. *Thomas Gray: Contemporary Essays*. Liverpool: Liverpool University Press.

Ingram, Alexander. 1816. *Concise System of Mathematics in Theory and Practice for the use of Schools, Private Students and Practical Men*, 8th ed. Edinburgh.

Jackson, George. 1816. *Jackson's New and improved system of mnemonics, or Two hours' study in the art of memory. Applied to figures, chronology, geography, statistics, history, systematic tables, poetry, prose, and to the common transactions of life. Rendered familiar to every capacity, and calculated for the use of schools, as well as for those who have attended lectures on this subject* ... London: Printed for G. Jackson.

Jacobs, Struan. 1991. *Science and British Liberalism: Locke, Bentham, Mill and Popper*. Aldershot: Avebury.

Johnson, Samuel. (1755) 1996. *A Dictionary of the English Language*. Edited by Anne McDermott. Cambridge: Cambridge University Press.

Johnson, Samuel. 1963. "No. 107. Tuesday, 13 November 1753." In *The Yale Edition of the Works of Samuel Johnson*, vol. 2: *The Idler and The Adventurer*, edited by John M. Bullitt, W. J. Bate, and L. F. Powell, 440–445. New Haven, CT: Yale University Press.

Johnson, Steven. 2002. *Emergence: The Connected Lives of Ants, Brains, Cities and Software*. New York: Scribner.

Jones, Chris. 1993. *Radical Sensibility Lectures and Ideas in the 1790s*. London: Routledge.

Jones, Edward Thomas. 1821. *The English system of balancing books: Being an improved method, applicable to every mode of book-keeping by double or single entry*. London: T. C. Hansard.

Jones, Steven. 1998. "Representing Rustics: Satire, Counter-Satire, and Emergent Romanticism." *Wordsworth Circle* 29 (1): 60–67.

Justman, Stewart. 1995. "Romanticism and the Rhetoric of Difference." *North Dakota Quarterly* 62 (3): 151–159.

Kant, Immanuel. 1970. "An Answer to the Question: 'What Is Enlightenment?'" In *Kant's Political Writings*, translated by H. B. Nisbet, edited by Hans Siegbert Reiss, 54–60. Cambridge: Cambridge University Press.

Kant, Immanuel. 1998. *Critique of Pure Reason*. Translated by P. Guyer and A. W. Wood. Cambridge: Cambridge University Press.

Kaoukji, N., and N. Jardine. 2010. "'A Frontispiece in Any Sense They Please?' On the Significance of the Engraved Title-Page of John Wilkins's *A Discourse concerning A NEW world & Another Planet, 1640*." *Word and Image* 26 (4): 429–447.

Kaufer, David S., and Kathleen M. Carley. 1993. *Communication at a Distance: The Influence of Print on Sociocutural Organization and Change, Communication*. Hillsdale, NJ: Erlbaum.

Kaufmann, David. 1995. *The Business of Common Life: Novels and Classical Economics between Revolution and Reform*. Baltimore, MD: The Johns Hopkins University Press.

Kaul, Suvir. 2000. *Poems of Nation, Anthems of Empire: English Verse in the Long Eighteenth Century*. Charlottesville: University Press of Virginia.

Keach, William. 1997. "Poetry, after 1740." In *The Cambridge History of Literary Criticism,* edited by H. B. Nisbet and Claude Rawson, 117–166. Cambridge: Cambridge University Press.

Keckermann, Bartholomäus. 1608. *Systema disciplinæ politicæ*. Hanoviæ: Apud Guilielmum Antonium.

Keckermann, Bartholomäus, and Peter Antonius. 1610. *Systema s.s. theologiæ, tribus libris adornatum*. Editio vltima. Hanoviæ: Apud Petrus Antonium.

Keckermann, Bartholomäus, and Wilhelm Antonius. n.d. *Systema logicae: tribus libris adornatum*. Editio secunda. Hanoviæ: Apud Guilielmum Antonium.

Keckermann, Bartholomäus, and Wilhelm Antonius. n.d. *Systema rhetoricae, in quo artis praecepta plenè & methodicè traduntur, & tota simul ratio studii Eloquentiae tam quo ad epistolas & colloquia familiaria*. Hanoviæ: Apud Guilielmum Antonium.

Keckermann, Bartholomäus, and Wilhelm Antonius Erben. n.d. *Systema geographicum duobus libris adornatum & publicè olim prælectum*. Hanoviæ: Apud Hæredes Guilielmi Antonii.

Keckermann, Bartholomäus, and Wilhelm Antonius Erben. n.d. *Systema systematum clarissimi viri Dn. Bartholomæi Keckermanni,: omnia huius autoris scripta philosophica vno volumine comprehensa lectori exhibens; idque duobus tomis quorum prior discipulum instrumentales … posterior ipsam Pædian philosophicam*. Hanoviæ: Apud Hæredes Guilielmi Antonii.

Keckermann, Bartholomäus, and Gemma. 1661. *Systema compendiosum totius mathematices hoc est geometriæ, opticæ, astronomiæ, et geographiæ publicis prælectionibus anno 1605 in celeberrimo Gymnasio Dantiscano propositum*. Oxonii: Excudebat Gulielmus Hall pro Francisco Oxlad. Early English Books Online.

Keckermann, Bartholomäus, Petrus Janichius, and Peter Antonius. n.d. *Systema compendiosum totius mathematices, hoc est, geometriæ, opticæ, astronomiæ, et geographiæ*. Hanoviæ: Apud Petrum Antonium.

Keckermann, Bartholomäus, and Georg Pauli. 1607. *Systema ethicæ, tribus libris adornatum & publicis prælectionibus traditum in Gymnasio Dantiscano Bartholomæo Keckermanno Dantiscano, philosophiæ ibidem professore*. Londini: Ex officina Nortoniana.

Keen, Paul. 1999. *The Crisis of Literature in the 1790s: Print Culture and the Public Sphere*. New York: Cambridge University Press.

Keen, Paul, ed. 2004. *Revolutions in Romantic Literature: An Anthology of Print Culture, 1780–1832*. Peterborough, Canada: Broadview.

Keener, Frederick M. 1983. *The Chain of Becoming: The Philosophical Tale, the Novel, and a Neglected Realism of the Enlightenment: Swift, Montesquieu, Voltaire, Johnson, and Austen*. New York: Columbia University Press.

Keller, Evelyn Fox. 1985. *Reflections on Gender and Science*. New Haven, CT: Yale University Press.

Kelley, Donald R. 1997. *History and the Disciplines: The Reclassification of Knowledge in Early Modern Europe*. Rochester, NY: University of Rochester Press.

Kelley, Donald R., and Richard H. Popkin, eds. 1991. *The Shapes of Knowledge from the Renaissance to Enlightenment*. Dordrecht: Kluwer Academic.

Kelly, Kevin. 1994. *Out of Control: The Rise of Neo-Biological Civilization*. Reading, MA: Addison-Wesley.

Kelly, Kevin. 2006. "Life, the Universe, and Everything: Interview with Seth Lloyd." *Wired*. Retrieved October 28, 2013. http://www.wired.com/2006/03/life-the-universe-and-everything/.

Kermode, Frank, John Hollander, Harold Bloom, Lionel Trilling, Martin Price, and J. B. Trapp, eds. 1973. *The Oxford Anthology of English Literature*. 2 vols. Oxford: Oxford University Press.

Kernan, Alvin. 1987. *Samuel Johnson and the Impact of Print*. Princeton, NJ: Princeton University Press.

King, Helen. 2007. *Midwifery, Obstetrics and the Rise of Gynaecology: The Uses of a Sixteenth-Century Compendium, Women and Gender in the Early Modern World*. Burlington, VT: Ashgate.

Knight, Isabel F. 1968. *The Geometric Spirit: The Abbé de Condillac and the French Enlightenment*. New Haven, CT: Yale University Press.

Kollmann, Augustus Frederic Christopher. 1806. *A new theory of musical harmony, according to a complete and natural system of that science*. London: Printed by W. Bulmer.

Kucich, Greg. 1997. "Romantic Studies: The State of the Art or Scholars in Search of a Period." *Wordsworth Circle* 28 (2): 82–84.

Laski, Harold J. 1936. *The Rise of European Liberalism*. London: George Allen & Unwin.

Laski, Harold J. 1997. *The Rise of English Liberalism*. New Brunswick, NJ: Transaction.

"Letter." 1777. *Gentleman's Magazine* 47: 105.

Levinson, Marjorie. 1986. *The Romantic Fragment Poem: A Critique of a Form*. Chapel Hill: University of North Carolina Press.

Lewis, W. S., ed. 1971. *The Yale Edition of Horace Walpole's Correspondence*. Vol. 25. New Haven, CT: Yale University Press.

Lincoln, Andrew. 1999. "What Was Published in 1798?" *European Romantic Review* 10 (2): 137–151.

Lipking, Lawrence. 1992. "Inventing the Eighteenth Centuries: A Long View." In *The Profession of Eighteenth-Century Literature: Reflections on an Institution*, edited by Leo Damrosch, 7–25. Madison: University of Wisconsin Press.

Lloyd, G. E. R. 2009. *Disciplines in the Making: Cross-Cultural Perspectives on Elites, Learning, and Innovation*. Oxford: Oxford University Press.

Lloyd, Seth. 2006. *Programming the Universe: A Quantum Computer Scientist Takes on the Cosmos*. New York: Knopf.

Locke, John. 1693. *Some thoughts concerning education*. London: Printed for A. and J. Churchill.

Lonsdale, Roger, ed. 1984. *The New Oxford Book of Eighteenth Century Verse.* Oxford: Oxford University Press.

Luhmann, Niklas. 1982. *The Differentiation of Society.* Translated by S. Holmes and C. Larmore. New York: Columbia University Press.

Luhmann, Niklas. 1997. "Globalization or World Society: How to Conceive of Modern Society?" *International Review of Sociology* 7 (1): 67–79.

Macaulay, Thomas Babington. (1831) 1965. "Parliamentary Reform." In *The English Reform Tradition 1790–1910*, edited by Sydney W. Jackman, 52–67. Englewood Cliffs, NJ: Prentice Hall..

Maclaurin, Colin. 1748. *An Account of Sir Isaac Newton's Philosophical Discoveries, in Four Books.* London: A. Millar and J. Nourse.

Malthus, Thomas Robert. 1798. *An essay on the principle of population, as it affects the future improvement of society: With remarks on the speculations of Mr. Godwin, M. Condorcet, and other writers.* London: Printed for J. Johnson.

Malthus, Thomas Robert. (1798/1803) 1992. *An essay on the principle of population, or, A view of its past and present effects on human happiness: with an inquiry into our prospects respecting the future removal or mitigation of the evils which it occasions.* Edited by Donald Winch. Cambridge: Cambridge University Press.

A Manual of the System of Discipline and Instruction for the Schools of the Public School Society of New York, Instituted in the Year 1805. 1845. New York: Egbert, Hovey & King.

Mandler, Peter. 1990. *Aristocratic Government in the Age of Reform: Whigs and Liberals, 1830–1852.* Oxford: Clarendon Press.

Marchand, Leslie A., ed. 1973–1994. *Byron's Letters and Journals.* 13 vols. London: John Murray.

Markman, Ellis. 1996. *The Politics of Sensibility: Race, Gender, and Commerce in the Sentimental Novel.* Cambridge: Cambridge University Press.

Martin, Benjamin. 1747. *Philosophia Britannica or A New and Comprehensive System of the Newtonian Philosophy, Astronomy and Geography.* 2 vols. London: C. Micklewright and Co.

Mavor, William Fordyce. 1785. *Universal stenography; or a new complete system of short writing Rendered perfectly easy to read & write; freed from all prolixity and obscurity; adapted to every purpose in which short writing is useful or ornamental and attainable in a few hours by the most common capacity: Being an improvement on the most celebrated systems that have been exhibited to the public for above a century past and superior to all in ease, elegance and expedition. Designed for the use of schools & private gentlemen*, 2nd ed. London: Printed for J. Harrison & Co. Eighteenth Century Collections Online (CB3327692300).

Mavor, William Fordyce. 1820. *Universal Stenography; or a Practical System of Short Hand*, 10th ed. London: Printed for Longman, Hurst, Rees, Orme, and Brown.

McCann, Andrew. 1998. *Cultural Politics in the 1790s: Literature, Activism, and the Public Sphere*. New York: St. Martin's Press.

McDowell, R. B. 1959. *British Conservatism, 1832–1914*. London: Faber & Faber.

McElroy, Davis D. 1952. "The Literary Clubs and Societies of Eighteenth Century Scotland and Their Influence on the Literary Productions of the Period from 1700 to 1800." PhD diss., University of Edinburgh. ProQuest (1771278444).

McElroy, Davis D. 1969a. *A Century of Scottish Clubs*. Edinburgh: Edinburgh Library.

McElroy, Davis D. 1969b. *Scotland's Age of Improvement: A Survey of Eighteenth-Century Literary Clubs and Societies*. Pullman: Washington State University Press.

McElroy, Lucille, and Davis D. McElroy. 1955. "Index to the Literary Clubs and Societies of Eighteenth Century Scotland." University of Edinburgh Library. Edinburgh Research Archive (1842/7312). http://hdl.handle.net/1842/7312.

McGann, Jerome J., ed. 1980–1993. *Lord Byron: The Complete Poetical Works*. 7 vols. Oxford: Clarendon Press.

McGann, Jerome J. 1983. *The Romantic Ideology: A Critical Investigation*. Chicago: University of Chicago Press.

McGann, Jerome J. 1985. "The Anachronism of George Crabbe." In *The Beauty of Inflections: Literary Investigations in Historical Method and Theory*, 294–312. Oxford: Clarendon Press.

McGann, Jerome J. 1994. "Canonade." *New Literary History* 25 (3): 487–504.

McGann, Jerome J. 1996. *The Poetics of Sensibility: A Revolution in Literary Style*. Oxford: Oxford University Press.

McKitterick, David. 2003. *Print, Manuscript, and the Search for Order, 1450–1830*. Cambridge: Cambridge University Press.

McLuhan, Marshall. 1962. *The Gutenberg Galaxy: The Making of Typographic Man*. Toronto: University of Toronto Press.

McLuhan, Marshall. (1967) 2005. *The Medium Is the Massage: An Inventory of Effects*. Berkeley: Gingko Press.

McLuhan, Marshall. 1994. *Understanding Media: The Extensions of Man*. Cambridge, MA: MIT Press.

McLuhan, Marshall. 2003. *Understanding Me: Lectures and Interviews*. Edited by Stephanie McLuhan, and David Staines. Cambridge, MA: MIT Press.

McLuhan, Marshall, and David Carson. 2003. *The Book of Probes*. Corte Madera, CA: Gingko Press.

McMahon, Darrin M. 2001. *Enemies of the Enlightenment*. Oxford: Oxford University Press.

McNally, David. 1988. *Political Economy and the Rise of Capitalism: A Reinterpretation*. Berkeley: University of California Press.

Mellor, Anne Kostelanetz. 2000. *Mothers of the Nation: Women's Political Writing in England, 1780–1830*. Bloomington: Indiana University Press.

Merquior, José G. 1991. *Liberalism Old and New*. Boston: Twayne.

Michael, Ian. 1987. *The Teaching of English from the Sixteenth Century to 1870*. Cambridge: Cambridge University Press.

Mill, James. 1796. *An examination of Jones's English system of book-keeping*. London: Vernor & Hood.

Mill, John Stuart. 1873. *Autobiography*. London: Longmans, Green, Reader, and Dyer.

Mitchell, Leslie. 2000. "'Where Clubs Are Trumps': Review of *British Clubs and Societies 1580–1800*, by Peter Clark." *Times Literary Supplement* (March 10).

Mitchell, Robert. 2006. "Beautiful and Orderly Systems: Adam Smith on the Aesthetics of Political Improvement." In *New Voices on Adam Smith*, ed. Leonidas Montes and Eric Schliesser, 61–86. New York: Routledge.

Mitchell, Robert. 2005. "Adam Smith and Coleridge on the Love of Systems." *Coleridge Bulletin* 25: 54–60.

Monthly Magazine, and British Register. 1797. London: Printed for R. Phillips.

Moore, Gordon E. 1965. "Cramming More Components onto Integrated Circuits." *Electronics* 38 (8): 114–117.

Morin, Edgar. 1992. "The Concept of System and the Paradigm of Complexity." In *Context and Complexity*, edited by Magoroh Maruyama, 125–138. New York: Springer.

Morowitz, Harold J. 2002. *The Emergence of Everything: How the World Became Complex*. Oxford: Oxford University Press.

Morrison, Charles. 1822. *A complete system of practical book-keeping, applicable to all kinds of business; Exemplified in four sets of books, of individual and partnership concerns*, 3rd ed. Glasgow: R. Chapman.

Morrison, James C. 1977. "Philosophy and History in Bacon." *Journal of the History of Ideas* 38 (4): 585–606.

Morse, David. 2000. *The Age of Virtue: British Culture from the Restoration to Romanticism*. New York: St. Martin's Press.

Myrone, Martin, and Lucy Peltz. 1999. *Producing the Past: Aspects of Antiquarian Culture and Practice, 1700–1850.* Aldershot: Ashgate.

Nemoianu, Virgil. 1989. *A Theory of the Secondary*. Baltimore, MD: The Johns Hopkins University Press.

Newton, Isaac. (1679) 1960. *The Correspondence of Isaac Newton*. Vol. 2. Edited by H. W. Turnbull. Cambridge: Cambridge University Press.

Newton, Isaac. 1687. *Philosophiæ naturalis principia mathematica.* Londini: Jussu Societatis Regiæ ac Typis Josephi Streater. Early English Books Online.

Newton, Isaac. (1687) 1999. *The Principia: Mathematical Principles of Natural Philosophy*. Translated by I. Bernard Cohen and Anne Miller Whitman. Berkeley: University of California Press.

Newton, Isaac. (1704) 1952. *Opticks; or, A Treatise of the Reflections, Refractions, Inflections & Colours of Light. Based on the 4th ed., London, 1730.* New York: Dover.

Newton, Issac. 1728. *A Treatise of the System of the World.* London: F. Fayram.

Newton, Isaac. (1728) 2004. *A Treatise of the System of the World.* Mineola, NY: Dover.

Newton, Isaac. 1978. *Unpublished Scientific Papers of Isaac Newton: A Selection from the Portsmouth Collection in the University Library, Cambridge.* Edited and translated by A. Rupert Hall and Marie Boas Hall. Cambridge: Cambridge University Press.

Nicholson, Andrew, ed. 1991. *Lord Byron: The Complete Miscellaneous Prose.* Oxford: Clarendon Press.

Nixon, Cheryl L., ed. 2008. *Novel Definitions: An Anthology of Commentary on the Novel 1688–1815.* Peterborough, Canada: Broadview Press.

Norris, Christopher. 2000. *Quantum Theory and the Flight from Realism: Philosophical Responses to Quantum Mechanics.* London: Routledge.

Nussbaum, Felicity A. 2005. "Biography and Autobiography." In *The Cambridge History of Literary Criticism*, vol. 4: *The Eighteenth Century*, edited by H. B. Nisbet and Claude Rawson, 302–315. Cambridge: Cambridge University Press.

Oerlemans, Onno. 2002. *Romanticism and the Materiality of Nature.* Toronto: University of Toronto Press.

Okabe, Atsuyuki, Barry Boots, Kokichi Sugihara, and Sung Nok Chiu. 2000. *Spatial Tessellations: Concepts and Applications of Voronoi Diagrams*, 2nd ed. Chichester, England: Wiley.

Oldys, William. 1747–1766. *Biographia Britannica: or, the lives of the most eminent persons who have flourished in Great Britain and Ireland … Digested in the manner of Mr. Bayle's … Dictionary. To which are added, a supplement and appendix …* 6 vols. London.

Olson, Richard. 1975. *Scottish Philosophy and British Physics, 1750–1880: A Study in the Foundations of the Victorian Scientific Style*. Princeton, NJ: Princeton University Press.

Olson, Richard. 1990. *Science Deified & Science Defied: The Historical Significance of Science in Western Culture*, vol. 2. 2 vols. Berkeley: University of California Press.

Olson, Richard. 1993. *The Emergence of the Social Sciences, 1642–1792*. New York: Twayne.

O'Neill, Michael. 1997. "The Lampless Deep: Criss-Crossing Strains in Current Romantic Criticism." *Romanticism on the Net* 7.

Ong, Walter J. 1956. "System, Space, and Intellect in Renaissance Symbolism." *Bibliothèque d'Humanisme et Renaissance* 18 (2): 222–239.

Ong, Walter J. 2004. *Ramus, Method, and the Decay of Dialogue: From the Art of Discourse to the Art of Reason*. Chicago: University of Chicago Press.

Orton, William Aylott. 1945. *The Liberal Tradition: A Study of the Social and Spiritual Conditions of Freedom*. Port Washington, NY: Kennikat Press.

Outram, Dorinda. 1995. *The Enlightenment*. Cambridge: Cambridge University Press.

Packham, Catherine. 2007. "Feigning Fictions: Imagination, Hypothesis, and Philosophical Writing in the Scottish Enlightenment." *Eighteenth Century* 48 (2): 149–171.

"Palgrave Pivot." 2012. Retrieved May 23, 2012 from http://www.palgrave.com/pivot/.

Papps, Thomas. 1818. *An Improved System of Book-Keeping Being the Introduction to Papp's Patent Improved Account Books*. London: T. C. Hansard.

Partridge, Eric. (1924) 1979. *Eighteenth Century English Romantic Poetry (Up till the publication of the "Lyrical Ballads," 1798)*. London: Norwood Editions.

Patterson, Annabel M. 1997. *Early Modern Liberalism*. Cambridge: Cambridge University Press.

Paul, John. 1776. *A system of the laws relative to bankruptcy*. London: W. Strahan & M. Woodfall.

Payne, John. 1791. *Universal geography formed into a new and entire system: Describing Asia, Africa, Europe, and America, with their subdivisions of empires, kingdoms, states, and republics*. 2 vols. London: Printed for J. Payne.

Pennecuik, Alexander. 1720. *Streams from Helicon: Or, Poems on Various Subjects. In three parts*, 2nd ed. London: Printed for the author.

Perkins, David. 1992. *Is Literary History Possible?* Baltimore, MD: The Johns Hopkins University Press.

Phelps, William Lyon. 1893. *The Beginnings of the English Romantic Movement: A Study in Eighteenth-Century Literature*. Boston: Ginn.

Phillipson, Nicholas T. 2010. *Adam Smith: An Enlightened Life*. London: Allen Lane.

Pitre, David Wayne. 1980. "Francis Jeffrey's Journal: A Critical Edition." PhD diss., University of South Carolina. ProQuest (303096055).

The Plain Dealer: being Select Essays on sever curious subjects … & other branches of Polite Literature. 1724.

Plant, Sadie. 1997. *Zeros and Ones: Digital Women and the New Technoculture*. London: Fourth Estate.

"*Playboy* Interview with Marshall McLuhan." 1969. *Playboy*, March. http://www.mcluhanmedia.com/m_mcl_inter_pb_01.html.

Plug, Jan. 2003. *Borders of a Lip: Romanticism, Language, History, Politics*. Albany: State University of New York Press.

Pomata, Gianna, and Nancy G. Siraisi, eds. 2005. *Historia: Empiricism and Erudition in Early Modern Europe*. Cambridge, MA: MIT Press.

Popkin, Richard H. 2003. *The History of Scepticism: From Savonarola to Bayle*. New York: Oxford University Press.

Porter, Lucy. n.d. "The Fair Intellectual Club." http://www.thefairintellectual-club.co.uk/, accessed January 7, 2016.

Porter, Roy. 2000. *The Creation of the Modern World: The British Enlightenment*. New York: Norton.

Porter, Roy. 2001. *The Enlightenment*. 2nd ed. Houndmills, England: Palgrave.

Pownall, Thomas. 1782. *A treatise on the study of antiquities as the commentary to historical learning, sketching out a general line of research: … With an appendix. … By T. Pownall*. London: Printed for J. Dodsley.

Quintana, Ricardo, and Alvin Whitley, eds. 1963. *English Poetry of the Mid and Late Eighteenth Century: An Historical Anthology*. New York: Knopf.

Rabinow, Paul, ed. 1984. *The Foucault Reader*. New York: Pantheon.

Radcliffe, David. 1993. *Forms of Reflection: Genre and Culture in Meditational Writing*. Baltimore, MD: The Johns Hopkins University Press.

Rae, Alice. 2009. "Extension." http://lightthroughmcluhan.org/extension.html.

Rajan, Tilottama, and Julia M. Wright, eds. 1998. *Romanticism, History, and the Possibilities of Genre: Re-forming Literature, 1789–1837*. Cambridge: Cambridge University Press.

Rajan, Tilottama. 2000. "System and Singularity from Herder to Hegel." *European Romantic Review* 11 (2): 137–149.

Rajchman, John. 1985. "Philosophy in America." In *Post-Analytic Philosophy*, edited by John Rajchman and Cornell West. New York: Columbia University Press.

Ramos, Adela. 2010. "The Emergence of Species: Animals, Humans, and Literary Form in Eighteenth-Century Britain." PhD diss., Columbia University. ProQuest (858865416).

Ramus, Petrus, and William Kempe. 1592. *The art of arithmeticke in whole numbers and fractions in a more readie and easie method then hitherto hath bene published*. London: By Richard Field for Robert Dextar dwelling in Paules Church yard at the signe of the brasen serpent. Early English Books Online.

Rasch, William, and Cary Wolfe. 1995. "Introduction: The Politics of Systems and Environments." *Cultural Critique* 30: 5–13.

Raven, James. 1992. *Judging New Wealth: Popular Publishing and Responses to Commerce in England, 1750–1800*. Oxford: Clarendon Press.

Raven, James. 2000. "Historical Introduction: The Novel Comes of Age." In *The English Novel, 1770–1829: A Bibliographical Survey of Prose Fiction Published in the British Isles*, edited by Peter Garside, James Raven, and Rainer Schöwerling, 15–110. Oxford: Oxford University Press.

Raylor, Timothy. 1994. *Cavaliers, Clubs, and Literary Culture: Sir John Mennes, James Smith, and the Order of the Fancy*. Newark: University of Delaware Press.

Reid, Ian. 2004. *Wordsworth and the Formation of English Studies*. Aldershot: Ashgate.

Reiman, Donald, ed. 1972. *The Romantics Reviewed: Part A, The Lake Poets*. 2 vols. New York: Garland.

Reiss, Hans, ed. 1970. *Kant's Political Writings*. Translated by H. B. Nisbet. Cambridge: Cambridge University Press.

"Review of *Biographical History of England*, by James Granger." 1769. *Monthly Review* 41:206–216.

"Review of *Essays Moral, Philosophical, and Political*, by John Mills." 1772. *Monthly Review* 46:382–383.

"Review of the *Waverly Novels* and *Tales of my Landlord*." 1832. *Edinburgh Review*, April, 61–79.

Renwick, W. L. 1963. *English Literature 1789–1815*. Oxford: Clarendon Press.

Rhodes, Neil, and Jonathan Sawday. 2000. *The Renaissance Computer: Knowledge Technology in the First Age of Print*. London: Routledge.

Riasanovsky, Nicholas Valentine. 1995. *The Emergence of Romanticism*. New York: Oxford University Press.

Rieder, John. 1997. "The Institutional Overdetermination of the Concept of Romanticism." *Yale Journal of Criticism* 10 (1): 145–163.

Robbins, Caroline. 1959. *The Eighteenth-Century Commonwealthman: Studies in the Transmission, Development and Circumstance of English Liberal Thought from the Restoration of Charles II until the War with the Thirteen Colonies*. Cambridge, MA: Harvard University Press.

Rogers, Pat, ed. 1993. *Johnson and Boswell in Scotland: A Journey to the Hebrides*. New Haven, CT: Yale University Press.

Rooney, Ellen. 1998. "Novel Times, or, the Imitation of Life." *Novel* 31 (3): 286–303.

Rosenblum, Nancy L. 1987. *Another Liberalism: Romanticism and the Reconstruction of Liberal Thought*. Cambridge, MA: Harvard University Press.

Rothstein, Eric. 1975. *Systems of Order and Inquiry in Later Eighteenth-Century Fiction*. Berkeley: University of California Press.

Rotunda, Ronald D. 1986. *The Politics of Language: Liberalism as Word and Symbol*. Iowa City: University of Iowa Press.

Roughley, Thomas. 1823. *The Jamaica planter's guide; or, A system for planting and managing a sugar estate, or other plantations in that island, and throughout the British West Indies in general, illustrated with interesting anecdotes*. London: Longman, Hurst, Rees, Orme, and Brown.

Rousseau, Jean-Jacques. 1891. *The Confessions of Jean-Jacques Rousseau newly translated into English*. 2 vols. Philadelphia: Lippincott.

Rudy, Seth. 2010. "The Circle of Arts, the Compass of Nature: Generic Change and the Pursuit of Complete Knowledge in Britain, 1620–1817." PhD diss., New York University. ProQuest (598303126).

Rudy, Seth. 2014. *Literature and Encyclopedism in Enlightenment Britain: The Pursuit of Complete Knowledge*. New York: Palgrave Macmillan.

Ruggiero, Guido de. (1927) 1959. *The History of European Liberalism*. Translated by R. G. Collingwood. Boston: Beacon Press.

Rundell, Maria Eliza. 1824. *A New System of Domestic Cookery; Formed Upon Principles of Economy: And Adapted to the Use of Private Families*. London: J. Murray.

Russell, Gillian. 2013. "'Who's Afraid for William Wordsworth?' Some Thoughts on 'Romanticism' in 2012." *Australian Humanities Review* 54: 66–80.

Russell, Gillian, and Clara Tuite, eds. 2002. *Romantic Sociability: Social Networks and Literary Culture in Britain 1770–1840*. Cambridge: Cambridge University Press.

Sanders, Andrew. 1994. *The Short Oxford History of English Literature*. Oxford: Clarendon Press.

Sarsi, Lothario [Orazio Grassi]. (1619) 1960. *The Astronomical and Philosophical Balance.* Translated by C. D. O'Malley. Philadelphia.

Schmidt, James, ed. 1996. *What Is Enlightenment? Eighteenth-Century Answers and Twentieth-Century Questions*. Berkeley: University of California Press.

Schmidt, James. 2011. "Misunderstanding the Question: 'What Is Enlightenment?' Venturi, Habermas, and Foucault." *History of European Ideas* 37 (1): 43–52.

Shaddy, Robert A. 2000. "Grangerizing: One of the Unfortunate Stages of Bibliomania." *Book Collector* 49 (4): 535–546.

Sharpe, Kevin, and Steven N. Zwicker, eds. 1998. *Refiguring Revolutions: Aesthetics and Politics from the English Revolution to the Romantic Revolution*. Berkeley: University of California Press.

Siegel, Joel G., and Jae K. Shim. 2005. *Dictionary of Accounting Terms*. 4th ed. Hauppauge, NY: Barron's.

Simpson, David. 1993. *Romanticism, Nationalism, and the Revolt against Theory*. Chicago: University of Chicago Press.

Siskin, Clifford. 1988. *The Historicity of Romantic Discourse*. New York: Oxford University Press.

Siskin, Clifford. 1997. "A Response to Richard Terry's 'Literature, Aesthetics, and Canonicity in the Eighteenth Century.'" *Eighteenth Century Life* 21 (1): 104–107.

Siskin, Clifford. 1998a. *The Work of Writing: Literature and Social Change in Britain, 1700–1830*. Baltimore, MD: The Johns Hopkins University Press.

Siskin, Clifford. 1998b. "The Year of the System." In *1798: The Year of Lyrical Ballads*, edited by Richard Cronin, 9–31. Basingstoke, England: Macmillan.

Siskin, Clifford. 2001a. "Novels and Systems." *Novel* 34 (2): 202–215.

Siskin, Clifford. 2001b. "VR Machine: Romanticism and the Physical." *European Romantic Review* 12 (2): 158–164.

Siskin, Clifford. 2005. "More Is Different: Literary Change in the Mid- and Late Eighteenth Century." In *The Cambridge History of English Literature, 1660–1780*, edited by John J. Richetti, 795–823. Cambridge: Cambridge University Press.

Siskin, Clifford. 2006. "William Wordsworth." In *Oxford Encyclopedia of British Literature*, edited by David Kastan. Oxford: Oxford University Press.

Siskin, Clifford. 2007. "Textual Culture in the History of the Real." *Textual Cultures* 2 (2): 118–130.

Siskin, Clifford. 2009a. "The Problem of Periodization: Enlightenment, Romanticism, and the Fate of System." In *The Cambridge History of English Romantic Literature*, edited by James Chandler, 101–126. Cambridge: Cambridge University Press.

Siskin, Clifford. 2009b. "Re-Mediating Ralph." *New Literary History* 40 (4): 719–727.

Siskin, Clifford. 2015. "The Rise of the 'Rise' of the Novel." In *Oxford History of the Novel in English*, Vol. 2: *English and British Fiction 1750–1820*, edited by Peter Garside, and Karen O'Brien, 615–630. Oxford: Oxford University Press.

Siskin, Clifford, and William Warner. 2008. "Stopping Cultural Studies." *Profession* 2008:94–107.

Siskin, Clifford, and William Warner, eds. 2010. *This Is Enlightenment.* Chicago: University of Chicago Press.

Sitter, John, ed. 2001. *The Cambridge Companion to Eighteenth-Century Poetry.* Cambridge: Cambridge University Press.

Skinner, Andrew. 2002. "Adam Smith: Science and System." Paper presented at Conference of The International Network for Economic Method, University of Stirling, September 1–2.

Skinner, Gillian. 1999. *Sensibility and Economics in the Novel, 1740–1800: The Price of a Tear.* Basingstoke, England: Macmillan.

Slowik, Edward. 2014. "Descartes' *Physics*." In *The Stanford Encyclopedia of Philosophy*, edited by Edward N. Zalta. Stanford, CA: The Metaphysics Research Lab, Center for the Study of Language and Information. http://plato.stanford.edu/entries/descartes-physics/.

Small, Helen. 2001. "'For the use of the Ladies': Review of *Bluestocking Feminism: Writings of the Bluestocking Circle, 1738–1785*." *Times Literary Supplement* (April 6).

Smith, Adam. (1759) 1976. *The Theory of Moral Sentiments.* Edited by D. D. Raphael and A. L. Macfie. Oxford: Clarendon Press.

Smith, Adam. (1776) 1976. *An Inquiry into the Nature and Causes of the Wealth of Nations.* Edited by R. H. Campbell and A. S. Skinner. Oxford: Clarendon Press.

Smith, Adam. (1795) 1980. *Essays on Philosophical Subjects. The Glasgow Edition of the Works and Correspondence of Adam Smith.* Edited by D. D. Raphael and A. S. Skinner. Oxford: Clarendon Press.

Smith, Adam. 1963. *Lectures on Rhetoric and Belles Lettres: Delivered in the University of Glasgow by Adam Smith Reported by a Student in 1762–63.* Edited by John M. Lothian. London: Thomas Nelson and Sons.

Smith, Adam. 1987. *The Correspondence of Adam Smith. The Glasgow Edition of the Works and Correspondence of Adam Smith*, 2nd ed. Edited by Ernest Campbell Mossner and Ian Simpson Ross. Oxford: Clarendon Press.

Smith, Dinitia. 2004. "Parlaying an Affinity for Austen into an Unexpected Best Seller." *New York Times*, June 14, E1, E5.

Smocovitis, Vassiliki Betty. 1996. *Unifying Biology*. Princeton, NJ: Princeton University Press.

Spender, Dale. 1986. *Mothers of the Novel: 100 Good Women Writers before Jane Austen*. London: Pandora.

St. Clair, William. 2004. *The Reading Nation in the Romantic Period*. Cambridge: Cambridge University Press.

Sterne, Laurence. 1967. *A Sentimental Journey through France and Italy*. Edited by Graham Petrie. Harmondsworth, England: Penguin.

Stevenson, David. 2001. *The Beggar's Benison: Sex Clubs of Enlightenment Scotland and Their Rituals*. East Linton, Scotland: Tuckwell Press.

St. John, Henry, Viscount Bolingbroke. 1752. *Letters on the Study and Use of History*. London: Printed for A. Millar.

Stow, David. 1833. *Infant training; a dialogue, explanatory of the system adopted in the model infant school in Glasgow*. Glasgow: Printed for William Collins.

Strydom, Piet. 2000. *Discourse and Knowledge: The Making of Enlightenment Sociology*. Liverpool: Liverpool University Press.

Sullivan, Alvin, ed. 1983. *British Literary Magazines: The Romantic Age, 1789–1836*. Westport, CT: Greenwood Press.

Sykes, Alan. 1997. *The Rise and Fall of British Liberalism 1776–1988*. London: Longman.

The Systematic, or Imaginary, Philosopher: A Comedy in Five Acts. 1800. London: Printed by S. Gosnell for Jordan Hookham. Eighteenth Century Collections Online.

Tarbuck, Derya Gürses. 2008. "Researching an Eighteenth-Century Women's Intellectual Club." *Center and Clark Newsletter* 48: 4–6.

Taylor, Mark C. 2001. *The Moment of Complexity: Emerging Network Culture*. Chicago: Chicago University Press.

Taylor, Martha. 1831. *The housekeeper's guide; or, a system of modern cookery …* Reading.

Taylor, Samuel. 1832. *Taylor's system of stenography, or short-hand writing. Revised and improved, after considerable practice, by John Henry Cooke*. London: William Crofts.

Taylor, William. 1798. "Review of *Observations sur le Sentiment du Beau et du Sublime*, by Emmanuel Kant, translated by Hercules Peyer-Imhoff." *Monthly Review* 25:584–85.

Terry, Richard, Jonathan Brody Kramnick, Howard D. Weinbrot, Barbara M. Benedict, Trevor Ross, Robert Crawford, J. Paul Hunter, and Thomas Bunnell. 1997a. "Forum." *Eighteenth Century Life* 21 (3): 79–99.

Terry, Richard, Thomas P. Miller, and Clifford Siskin. 1997b. "Forum." *Eighteenth Century Life* 21 (1): 80–107.

Todd, Janet, ed. 1984. *Dictionary of British and American Women Writers 1660–1800.* Totowa, NJ: Rowman & Allanheld.

Todd, Janet. 1986. *Sensibility: An Introduction.* London: Methuen.

Tolley, Michael J. 1998. "Preromanticism." In *A Companion to Romanticism*, edited by Duncan Wu, 12–22. Oxford: Blackwell.

Tresch, John. 2012. *The Romantic Machine: Utopian Science and Technology after Napoleon.* Chicago: University of Chicago Press.

Trotter, James. 1843. *A Key to Ingram's Concise System of Mathematics; Containing Solutions of All the Questions Prescribed in That Work*, 4th ed. Edinburgh.

Tuite, Clara. 2002. *Romantic Austen: Sexual Politics and the Literary Canon.* Cambridge: Cambridge University Press.

Turnbull, H. W., ed. 1960. *Correspondence of Isaac Newton.* 3 vols. Vol. 2. Cambridge: Cambridge University Press.

Twain, Mark. (1883) 1999. *Life on the Mississippi.* Documenting the American South (Project), and University of North Carolina at Chapel Hill Library. Chapel Hill: Academic Affairs Library, University of North Carolina at Chapel Hill. http://docsouth.unc.edu/southlit/twainlife/menu.html.

Tytler, Alexander Fraser, Lord Woodhouselee. 1782. *Plan and outlines of a course of lectures on Universal History, ancient and modern.* Edinburgh: Printed for W. Creech.

Unger, Roberto Mangabeira. 1975. *Knowledge and Politics.* New York: Free Press.

The Unsex'd Females, a Poem. 1798. London: Cadell and Davies.

"Up Front." 2006. *New York Times Book Review*, March 5.

Venn, Henry. 1759. *The complete duty of man: or, a system of doctrinal and practical Christianity. To which are added forms of prayer and offices of devotion for the various circumstances of life. Designed for the Use of Families*, 3rd ed. London.

Wall, Cynthia. 2006. *The Prose of Things: Transformations of Description in the Eighteenth Century.* Chicago: University of Chicago Press.

Wallerstein, Immanuel Maurice. 2004. *World-systems Analysis: An Introduction.* Durham: Duke University Press.

Walpole, Horace. 1849. *Anecdotes of painting in England: with some account of the principal artists; and incidental notes on other arts; also a catalogue of engravers who have been born or resided in England.* New ed. 3 vols. London: Henry G. Bohn.

Walpole, Horace. 1937–1980. *The Complete Correspondence of Horace Walpole.* 48 vols. Edited by W. S. Lewis. New Haven, CT: Yale University Press.

"Walter Scott—Has History Gained by His Writings?" 1847. *Fraser's Magazine* 36: 345–351.

Warner, Richard. 1828. *Sunday-evening discourses: or, A compendious system of scriptural divinity, for the use of households: To which is added a visitation sermon*. 2 vols. London: Longman, Rees, Orme, Brown & Green.

Warner, William B. 1998. *Licensing Entertainment: The Elevation of Novel Reading in Britain, 1684–1750*. Berkeley: University of California Press.

Watson, J. R., ed. 1989. *Pre-Romanticism in English Poetry of the Eighteenth Century: The Poetic Art and Significance of Thomas Gray, Collins, Goldsmith, Cowper, Crabbe, A Casebook*. London: Macmillan.

Watts, Isaac. 1741. *The Improvement of the Mind: or, A Supplement to the Art of Logick: Containing a Variety of Remarks and Rules for the Attainment and Communication of Useful Knowledge, in Religion, in the Sciences, and in Common Life*. London: J. Brackstone.

"Waverley: Supposed by W. Scott." 1814. In *The British Critic*, 189–211. London: F. C. and J. Rivington.

Weinberger, David. 2011. *Too Big to Know: Rethinking Knowledge Now That the Facts Aren't the Facts, Experts Are Everywhere, and the Smartest Person in the Room Is the Room*. New York: Basic Books.

Whale, John C. 2000. *Imagination under Pressure, 1789–1832: Aesthetics, Politics, and Utility*. Cambridge: Cambridge University Press.

White, Gilbert. 1789. *The Natural History and Antiquities of Selborne, in the County of Southampton: With Engravings, and an Appendix*. London: Printed by T. Bensley for B. White and Son.

Willard, Charles Arthur. 1996. *Liberalism and the Problem of Knowledge: A New Rhetoric for Modern Democracy*. Chicago: University of Chicago Press.

Williams, Hank. 2008. "The Death of the Relational Database." *Why Does Everything Suck?* (February 5). http://whydoeseverythingsuck.com/2008/02/death-of-relational-database.html.

Williams, Raymond. 1958. *Culture and Society, 1780–1950*. London: Chatto & Windus.

Williams, Raymond. 1961. *The Long Revolution: An Analysis of the Democratic, Industrial, and Cultural Changes Transforming Our Society*. New York: Columbia University Press.

Williams, Raymond. 1973. *The Country and the City*. New York: Oxford University Press.

Williams, Raymond. 1976. *Keywords: A Vocabulary of Culture and Society*. New York: Oxford University Press.

Williams, Raymond. 1979. *Politics and Letters: Interviews with New Left Review*. London: New Left Books.

Williams, Raymond. 1983. *Writing in Society*. London: Verso.

Wilson, Edward O. 1999. *Consilience: The Unity of Knowledge*. London: Abacus.

Wise, Joseph. 1781. *The System. A Poem. In Five Books*. London: R. Faulder.

Wo, Ching-ling. 2004. "Re-orienting the British Enlightenment." PhD diss., State University of New York at Stony Brook. ProQuest (305056562).

Wolfram, Stephen. 2001. *A New Kind of Science*. Champaign, IL: Wolfram Media.

Woodhouselee, Alexander Fraser Tytler. 1782. *Plan and outlines of a course of lectures on universal history, ancient and modern, delivered in the University of Edinburgh, by Alexander Tytler, … Illustrated with maps*. Edinburgh: Printed for William Creech.

Woodman, Thomas, ed. 1998. *Early Romantics: Perspectives in British Poetry from Pope to Wordsworth*. London: Macmillan.

Woof, Robert, ed. 2001. *William Wordsworth: The Critical Heritage*, vol. 1. London: Routledge.

Woolfson, Adrian. 2002. "How Did the Slime Mould Cross the Maze? Review of *Emergence: The Connected Lives of Ants, Brains, Cities and Software*, by Steven Johnson, and *The Moment of Complexity: Emerging Network Culture*, by Mark Taylor." *London Review of Books* 24 (6): 27–28.

Wordsworth, William. (1798) 1965. *Lyrical Ballads*. Edited by R. L. Brett and A. R. Jones. London: Methuen.

Wordsworth, William. 1879. *Poems of Wordsworth*. Edited by Matthew Arnold. London: Macmillan.

Wordsworth, William. 1941–1949. *The Poetical Works of William Wordsworth*. 5 vols. Edited by Ernest de Selincourt. Oxford: Clarendon Press.

Wordsworth, William. 1974. *The Prose Works of William Wordsworth*. 3 vols. Edited by W. J. B. Owen and Jane Worthington Smyser. Oxford: Clarendon Press.

Wordsworth, William, and Dorothy Wordsworth. 1967. *The Letters of William and Dorothy Wordsworth: The Early Years, 1787–1805*. 2 vols. Edited by Ernest de Selincourt. Oxford: Oxford University Press.

Yates, Frances Amelia. 1966. *The Art of Memory*. Chicago: University of Chicago Press.

Yeo, Richard R. 2001. *Encyclopaedic Visions: Scientific Dictionaries and Enlightenment Culture*. Cambridge: Cambridge University Press.

Yoder, Laura E. 2012. "Evidence of Things Unreal: Collections and Writing of Horace Walpole." Master's thesis, New York University.

Young, Edward. 1966. *Conjectures on Original Composition (1759), A Scolar Press Facsimile*. Leeds: Scolar Press.

Zenil, Hector. 2011. "The World Is Either Algorithmic or Mostly Random." http://www.mathrix.org/liquid/archives/fqxi, accessed October 27, 2013.

Zenil, Hector. 2013. *A Computable Universe: Understanding and Exploring Nature as Computation*. Hackensack, NJ: World Scientific. Kindle edition.

INDEX

Page numbers in italics refer to illustrations.